Oceanography for Landlocked Classrooms

Monograph V

About the Editors

Gerry M. Madrazo, Jr. has been a senior director of instruction, science curriculum specialist, biology teacher, consultant and author. He has taught oceanography and marine biology, received presidential awards from the Mid-Atlantic Marine Education Association and was named Marine Educator of the Year by the University of North Carolina Sea Grant Program. He currently serves on the NABT Board of Directors. Madrazo received a B.S. in biology from the Ateneo de Zamboanga in the Philippines and his master's and Ph.D. in curriculum and instruction/science education from the University of North Carolina at Chapel Hill. He has done graduate work in botany at the University of Hawaii and the University of Wisconsin at Madison.

Paul B. Hounshell is a professor in the School of Education, University of North Carolina at Chapel Hill. He has been with the university since 1966 and previously taught biology and chemistry in public schools and was science supervisor for six years in Winston-Salem, North Carolina, and Charlotte, North Carolina. The author of books and articles about science education, he has a B.S. and M.S. in science education from the University of Virginia and a doctorate in education from UNC-Chapel Hill.

Acknowledgments

The production of this monograph would not have been possible without the help of various individuals, including the numerous authors who submitted manuscripts. It is a unique project in that it attempts to bring marine biology to landlocked classrooms. We are indebted to former NABT Board member **Ed Frazier**, NABT Executive Director **Patricia McWethy** and the NABT Board of Directors for their encouragements and foresights.

The editors are grateful to our Editorial Review Board, which reviewed all the manuscripts for this monograph: **Dirk Frankenberg**, University of North Carolina, Chapel Hill; **Frances L. Lawrence**, College of William & Mary, Virginia Institute of Marine Science, Gloucester Point, Virginia; **R. Dean Medley**, Center of Excellence, Eastern High School, Gibsonville, North Carolina; **Lindy Millman**, Oceanic Society/San Francisco Bay Chapter; **Gordon Plumblee**, Western Alamance High School, Elon College, North Carolina; and **Lundie Spence**, UNC Sea Grant Program, Raleigh, North Carolina. In addition to reviewing manuscripts, Lundie Spence also provided many valuable suggestions and ideas for the project. Many thanks to **Karen Jarrett** for typing and retyping the manuscripts on our word processor and to **Michelle Robbins**, director of publications and marketing at NABT, for working on this project. To these individuals and the NABT staff involved in this monograph's publication, our gratitude and appreciation.

Gerry M. Madrazo, Jr.
Paul B. Hounshell
Editors

Table of Contents

Authors

Steve K. Alexander, Department of Biology, University of Mary Hardin-Baylor, Belton, Texas 76513

Teresa Auldridge, science education consultant, Route 5, Box 185, Amelia, Virginia 23002

Norman D. Anderson, professor of science education, Department of Mathematics and Science Education, North Carolina State University, Raleigh, North Carolina 27695-7801

Nancy Chartier Balcolm, Connecticut Sea Grant, Marine Advisory Program, University of Connecticut Cooperative Extension Service, 43 Marne, Hamden, Connecticut 06514

Vicki Price Clark, program specialist, Mathematics & Science Center, 2401 Hartman Street, Richmond, Virginia 23223

Gregory J. Conway, Science Department, Highland Regional High School, Erial Road, Blackwood, New Jersey 08102

Nancy S. Cowal, media coordinator, Cape Hatteras School, P.O. Box 948, Buxton, North Carolina 27920

Rosanne W. Fortner, associate professor, School of Natural Resources, Ohio State University, 2021 Coffey Road, Columbus, OH 43210-1085

Barry W. Fox, extension specialist, 4-H marine / aquatic education, Virginia State University, Petersburg, Virginia 23803

William R. Hall, Jr., marine education specialist, University of Delaware, Sea Grant Program, 700 Pilottown Road, Lewes, Delaware 19958-1298

Paul B. Hounshell, professor of education, University of North Carolina, Chapel Hill, North Carolina 27514

Mary Ann Johnson, Science Department chairman, Weaver Education Center, 300 S. Spring St., Greensboro, North Carolina 27401

E. Barbara Klemm, co-director, Hawaii Marine Science Studies Project, University of Hawaii at Manoa, College of Education, Curriculum Research and Development Group, 1776 University Avenue, Honolulu, Hawaii 96822

Frances L. Lawrence, Sea Grant Marine Advisory Service, Virginia Institute of Marine Science, College of William and Mary, Gloucester Point, Virginia 23062

Gerry M. Madrazo, Jr., senior director of curriculum and instruction/science curriculum specialist, Guilford County Schools, 120 Franklin Boulevard, Greensboro, North Carolina 27401

L.W. McLamb, environmental science instructor, Salem High School, Virginia Beach, Virginia 23464

James O'Connor, associate professor of geoscience and city geologist, Department of Environmental Science, MB 44-04, University of the District of Columbia, Washington, DC 20008

Kevin T. Patton, Department of Life Science, St. Charles County Community College/Department of Physiology, St. Louis University Medical Center, 200 North Main Street, O'Fallon, Missouri 63366

John R. Sode, College of Education, Department of Curriculum and Instruction, University of Missouri–Columbia, 107 Townsend Hall, Columbia, Missouri 65211

Lundie Spence, marine education specialist, University of North Carolina–Sea Grant College Program, Box 8605, North Carolina State University, Raleigh, NC 27695-8605

J. Garrett Tomlinson, program specialist–Natural Science Center, Greensboro Public Schools, 4301 Lawndale Dr., Greensboro, North Carolina 27408

Karen Travers, program coordinator, Delaware Nature Society, Box 700, Hockessin, Delaware 19707

David A. Wright, Northeast Magnet School of Science and Engineering and the Visual Arts, 1847 N. Chautaqua Drive, Wichita, Kansas 67214

Emmett L. Wright, College of Teacher Education, Office of the Associate Dean, 017 Bluemont Hall, Kansas State University, Manhattan, Kansas 66506

Section 1. Getting Started

Why Marine Education?

Paul B. Hounshell
Gerry M. Madrazo, Jr.

If the oceans should die—it would be the final as well as the greatest catastrophe in the troublous story of man and the other animals and plants with whom man shares this planet.

Jacques-Yves Cousteau 1963

We may well be on the verge of causing the slow death of the world's oceans and seas, creating havoc for civilized man in the process. The trend surely can be reversed but it will take a monumental effort by men and women of all nations and it must begin soon. We do not know the answers and probably not even all the questions, but one thing is clear: saving the oceans will take understanding, world-wide cooperation and money.

Education is the key in this process. There can be no understanding without education and no cooperation without understanding; without either there will be little, if any, financial commitment.

Oceans cover more than 70 percent of the earth's surface and contain some 3 x 10 cubic miles of liquid water. Eighteen of the world's 20 largest cities have direct access to the ocean (Ingmanson & Wallace 1985). Of America's 50 states, 23 have ocean shoreline and a high percentage of our population lives in those 23 states.

What makes oceans so important? Why should oceanography be included in the science curriculum regardless of where we live? Let's examine some reasons.

Climate

Weather is an intricate, complex phenomena but "the driving force behind weather the world over is heat energy" (Ingmanson & Wallace 1985). Since oceans comprise more than 70 percent of the earth's surface, the majority of solar energy reaching the earth affects ocean waters. The atmosphere and the hydrosphere are in constant contact. "In fact, the uppermost 3 m of the ocean contain the same quantity of heat as the entire atmosphere. The ocean is the great modifier of temperature, moving massive amounts of heat slowly from place to place" (Ingmanson & Wallace 1985). Thus, oceans have a tremendous impact on climate in the U.S. and every other nation on the globe.

Young people and adults ask, "Why learn about the weather? We can't do anything about it." At this time we cannot "control" the weather but we can control how we react to and prepare for weather and climate. "Human beings are weather-sensitive creatures. Our industry, agriculture, sport, leisure, commerce, transportation, even our dispositions are all affected by weather." (Ingmanson & Wallace 1985). With weather–as with the other factors in our "list"–the more we know and understand, the more likely we are to make sound, rational decisions.

Recreation

The history of man, from earliest records to present day, is closely related to the sea. "Since the time of Homer, Plato and Aristotle, and probably long before, poets and philosophers have reflected on the sea" (Ingmanson & Wallace 1985). The sea has provided food and an important mode of transportation. Most of the world's major cities have been located alongside the sea; dominance of the oceans brought worldwide power. The oceans have played a significant role in human history and will be an important factor in the future.

"The sea has always challenged the minds and imagination of men" (Carson 1961). Think of the thousands of pages written by poets and philosophers, mariners and explorers, fishermen or just plain lovers of the sea. "The thought of the sea alone can be motivation for both young and adult." (Madrazo & Hounshell 1980). The sea has been a source of fun and dreams for generations of humans the world over. We swim, scuba dive, fish, sail, water ski and surf. In America alone, some 100 million people participate in recreational marine activities each year (Thurman 1987). The seas of the world have offered recreational opportunities–active or passive–for hundreds, probably thousands, of years.

Dumping

Despite efforts such as the 1972 International Ocean Dumping Agreement, seas have continued to be "garbage cans" for human trash, and the practice of ocean dumping continues unabated. Every bi-product of man–from sewage to radioactive waste–has been deposited in the sea: 9 million tons of solid waste and billions of gallons of sewage off America's shores every year. Worldwide, thousands of pounds of heavy metals are released into the sea yearly, along with unknown quantities of organics (DDT, 2. 4-D and many, many others) and well over 6 million metric tons of oil (not including spills). Using oceans as dump sites clearly creates a multitude of interrelated environmental problems; as human populations increase, the situation will only get worse.

Despite efforts such as the 1972 International Ocean Dumping Agreement, seas have continued to be "garbage cans" for human trash, and the practice of ocean dumping continues unabated.

The Economy

The oceans of the world act as aquatic roadways for international commerce. The shipping industry carries millions of tons of cargo all over the world. Probably half the cargo entering the U.S. comes by sea and is unloaded at our ports. As technology allows more efficient ships to be built, the price of per-ton shipping decreases and the demand increases. Air transportation is faster but nothing can match the low cost of ocean transportation. For now, at least, its future looks secure.

The seas have marked the key to power, prestige and military superiority since long before recorded history. The Greeks, the Romans, the Spanish, the British–and in World War II, the Americans–all have exerted their power through the superiority of their navies. Even with the advent of nuclear weapons, rocketry and laser technology, the seas remain a very viable, important factor in the worldwide balance of military power.

The commercial fishing industry, which provides food, oil and fertilizer for human use, harvests 71.3 million metric tons a year, almost 90 percent of that from ocean waters. The catch ranges from herring, sardines and anchovies to shrimp, squid, flounder and tuna. In addition, the harvest includes marine mammals such as whales and seals, invertebrates such as oysters and abalone, reptiles such as turtles and crocodiles, and numerous species of birds (Ingmanson & Wallace 1985).

The seas of the world contain atoms and molecules of nearly every element known to man. In addition, seafloor muds (rich in minerals), nodules (manganese) in deeper water, and gas and oil deposits all can and do provide useful products and materials. Salt (NaCl), iron, manganese, copper, cobalt, nickel, zinc, lead, silver and gold now are either being mined from seawater or are being considered for it. As marine mining technologies develop, many other mineral resources will be viable products for commercial interests (Ingmanson & Wallace 1985).

Farming in the marine environment (mariculture) is another frontier being investigated and, to some extent, used by man. "Currently, many species of fish (plaice, salmon, milkfish and tilapia), invertebrates (shrimp, abalone, mussels and oysters) and seaweeds are being grown commercially" (Ingmanson & Wallace 1985). Mariculture attempts to artificially duplicate an environment or grow an organism more efficiently in its natural environment (Lerman 1986). Advances in technology could enable mariculture to help ease the world's food crisis.

Future Technology

Potable water is an absolute necessity for human existence but its availability to much of the world's population is in question. The sea could well be a valuable source of fresh water for future generations, but the technology now available is too expensive to be feasible. The concern over the continued availability of water for humans and agriculture may diminish as technology is developed that removes salt and the multitude of other compounds and elements found in seawater.

Scientists also are looking at ways to harness some of the oceans' energy. The amount of kinetic energy stored in a wave is incredible. There are working plants that generate electricity through the rise and fall of tidal water and in some parts of the world tides vary as much as 49 feet. "A wave with a height of 1.3 m (4.3 ft.) breaking over a kilometer of coastline releases about 22,000 hp" (Ingmanson & Wallace 1985). With proper technology, wave energy could generate a significant amount of electricity that, like solar energy, is completely "clean."

Oceans & Society

People the world over are linked to the sea whether or not they live near the seashore. Decisions made about recreation, shipping, dumping and fisheries will have financial, sociological and environmental implications.

It will take a concerted effort to bring about change in a democratic society. To be able to make a difference, people must know, understand and appreciate how the marine environment works and what the results of human interaction will be. It is through education that the public will be prepared to make decisions about the oceans' future.

Perhaps John Culliney expressed it best when he wrote, "The oceans are the planet's last great living wilderness, man's only remaining frontier on earth, and perhaps his last chance to prove himself a rational species" (Culliney 1979).

References

Andrews, W.A. (1987). *Investigating aquatic ecosystems.* Englewood Cliffs, NJ: Prentice Hall.

Carson, R.L. (1961). *The sea around us.* New York: Oxford University Press.

Cheryl, C. (Project Wild Director). (1987). *Aquatic Project Wild.* Boulder, CO: Weston Regional Environmental Education Council.

Cousteau, J. Y. (1963). *The living sea.* New York: Harper and Row.

Culliney, J.L. (1979). *The forests of the sea: Life and death on the continental shelf.* New York: Anchor Press/Doubleday.

Ingmanson, D.E. & Wallace, W.J. (1985). *Oceanography: An introduction.* Belmont, CA: Wadsworth Publishing Co.

Kaufman, W. & Pilkey, O.H. Jr. (1983). *The beaches are moving; The drowning of America's shoreline.* Durham, NC: Duke University Press.

Lerman, M. (1986). *Marine biology environment, diversity and ecology.* Menlo Park, CA: The Benjamin/Cummings Publishing Co.

Madrazo, G. & Hounshell, P.B. (1980). Marine education in a land-based curriculum. *School Science and Mathematics, LXXX*(5), 363-370.

McConnaughey, B.H. (1978). *Introduction to marine biology.* St. Louis: C.V. Mosby Co.

Reseck, J. (1979). *Marine biology.* Reston, VA: Reston Publishing Co.

Thurman, H.V. (1987). *Essentials of oceanography* (2nd. ed.). Columbus, OH: Merrill Publishing Co.

An Action Agenda for Aquatic Education

Rosanne W. Fortner

Given the need for aquatic education (Goodwin & Schaadt 1978), the desire to teach about the world of water and the openly enthusiastic audience of students, what more could a teacher need for successful marine education? An ocean or a Great Lake would be nice, but no one pretends that those amenities are genuinely necessary (Fortner 1980). The bottom line is how to make marine education happen in the individual classroom, wherever it may be.

There have been attempts in the past to generate state-level support for marine education (Lanier 1988). By 1980, the National Marine Educators' Association (NMEA) had convinced 30 state education departments to establish a marine education coordinator position. In most cases this was a new "hat" given to the already busy science or environmental education coordinator, and few new initiatives were begun where none had existed before. Some coastal states used the position to advantage and benefited greatly from advocacy on the state level. Most participating states still recognize the title and can identify the individual currently responsible for marine education concerns.

A 1987 Educational Resources Information Center (ERIC) survey of state environmental education programs indicated that 41 percent of states include marine and freshwater education in the elementary curriculum, 51 percent in the secondary curriculum (Disinger 1987). The level of inclusion was not specified. While none of the states require aquatic education, a consultation with the state department of education may be a reasonable starting place. If there are existing efforts, they may serve as models. The teacher could at least determine who in the department makes curricular recommendations as the designated coordinator.

Other government agencies can also be consulted. The U.S. Fish and Wildlife Service and state wildlife agencies can be credited with state level support for aquatic education, especially as facilitated through Wallop-Breaux funds and active encouragement of *Project WILD - Aquatic* (Western Regional Environmental Education Council 1986).

The more common approach to inland marine education is for individual teachers to undertake the project themselves. Determined teachers will seek out experts as guest speakers, use the library to gather information for teaching and develop ideas for activities. Excitement for the subject matter is quite pervasive and may become a uniting force for interdisciplinary efforts.

Unfortunately, many of these valiant efforts duplicate what has already been done. This should not be necessary. Classroom-ready curriculum materials are available (e.g., Regan 1982; Marine education, et al. 1988), as are community interest, a literature database (Sea Grant Marine Education Center) and a professional support organization–the National Marine Educators' Association. A plan has even been developed for enhancing inland aquatic education (Crane, et al. 1988). It is based on a need to provide education about the Great Lakes to schools of that region, but it offers a list of ideas that are easily adapted to marine education in inland schools.

Model for Inland Aquatic Education

In 1987 Great Lakes education was deemed a priority issue by the Great Lakes Commission, which serves as a regional voice for issues common to the Great Lakes states. The commission recognized that "the key to better resource management ultimately rests with an active, informed public, an ethic that must be instilled in formal K-12 classroom training" (Crane, et al. 1988). A task force established in the eight states and one province of the region consisted of two types of educators–K-12 classroom and those from museums, parks and zoos–along with government agencies and stakeholding interest groups. The group organized a series of six regional roundtable meetings to consider ways obstacles could be turned into opportunities. To a large extent, the issues considered and the strategies adopted are identical to those that would apply to inland marine education anywhere. They are presented here as a model that may assist others. (Credit for compiling these suggestions goes to Tom Crane, natural resources specialist, and Michael Donahue, executive director of the Great Lakes Commission.) How many of the following recommendations address your own needs as an inland marine educator?

Recommendations

The report recommended:

1) The development of a clearinghouse for centralized access to curricula, publications, AV materials and adaptable materials from other regions.

For marine teaching aids, the Marine Education Materials System (MEMS) functioned as a clearinghouse in the 1980s, amassing a huge inventory of print materials on microfiche. Major university libraries in every state received MEMS collections and still maintain them on file. Federal budget reallocations have terminated the activity of MEMS on a national level, but the same kinds of items have continued to be catalogued by the Educational Resources Information Center/Science, Mathematics and Environmental Acquisitions Clearinghouse (ERIC/SMEAC). A search using the key words "marine education, oceanography" and the grade range or geographic area can result in a useful list of materials. The materials must still be printed from microfiche or ordered. The best option would seem to be local access for educators to relevant materials, probably through a school system's bureau of teaching materials. A regional or national center would coordinate evaluation and spread of information to those bureaus, as libraries now share information.

2) a Speaker's Bureau with an annually updated directory.

The same local clearinghouse should maintain a database of experts willing to visit classrooms, assist with science projects and provide teacher training and other services.

3) an Educators' Network to provide ongoing contact among teachers who have special regional interests in aquatic education.

The level of formality, activities and means of communication should be decided among the group to facilitate the widest possible participation of teachers. For marine and aquatic education, such groups already exist. The NMEA, in addition to its annual meeting, newsletter and journal for almost 1300 members, has 14 state and regional chapters with similar means of ongoing contact. You need never be far from another marine educator; the regional liaison is an especially valuable one for

Table 1. National Marine Educators' Association and regional chapters.

National Marine Educators' Association

P.O. Box 51215, Pacific Grove, CA 93950

Chapters and constituent states

Consortium of Aquatic and Marine Educators of Ohio (CAMEO)
Florida Marine Science Educators' Association (FMSEA)
Georgia Association of Marine Educators (GAME)
Gulf of Maine Marine Education Association (GOMMEA–ME, NH, VT, Maritimes)
Massachusetts Marine Educators (MME)
Mid-Atlantic Marine Education Association (MAMEA–NC, VA, DC, MD, DE)
New Jersey Marine Education Association
New York State Marine Education Association (NYSMEA)
Northwest Association of Marine Educators (NAME–WA, OR, AK, BC)
OCEANIA (Hawaii, Guam, Samoa, New Zealand, Australia, Fiji, Nui, other Pacific nations)
South Carolina Marine Educators Association (SCMEA)
Southern Association of Marine Educators (SAME–LA, MS, AL)
Southwest Marine Educators Association (CA, NV, AZ)
Southeastern New England Marine Educators (SENEME–RI, CT)
Texas Marine Education Association (TMEA)

learning about new creative approaches and adaptable activities for the area. Table 1 lists regional NMEA chapters.

Other Ideas

Many teachers are unaware of or have little access to existing materials (Seager 1988). They often do not have the time or resources to adapt materials that might be available, and they may be unaware that MAE's scope allows its infusion into all curricular areas. The scope and sequence of the curricula must be examined to identify appropriate infusion points and to be certain that materials exist for use in those areas.

Specifically, there is need for a comprehensive inventory of available materials. Texas A&M publishes a directory of Sea Grant Education Materials from all U.S. Sea Grant programs (1988). There is also a *Directory of Great Lakes Education Materials* (Cole-Misch 1987) and a *Directory of Marine Education Resources* (Regan 1982), as well as local listings from most NMEA chapters. Keeping these directories available and updated for teachers should be a priority of the clearinghouses or organizations. The listings will also help to identify gaps that curriculum developers can fill, particularly in relation to current aquatic issues.

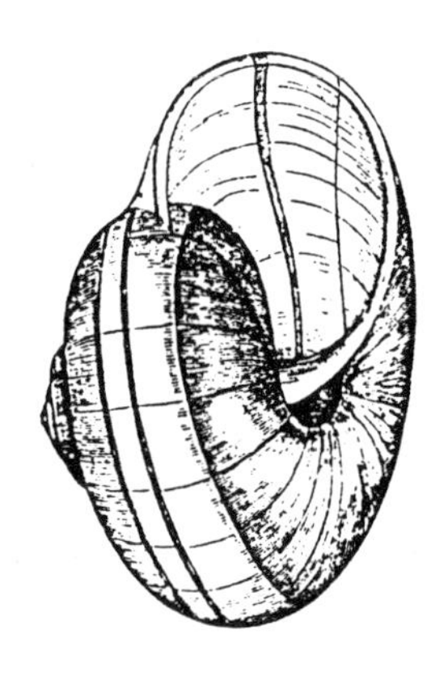

There is also a need for expanded inservice and pre-service teacher training for marine education. Local educational institutions should be contacted to see if they might teach continuing education courses for teachers. Seminar series by science organizations or college departments may also be available. Some form of credit to teachers for participation is an important consideration.

Existing materials also should be evaluated because of their great differences in sophistication, depth, breadth and quality. Characterizing them in terms of broad criteria can help teachers assess relative values and expend limited funds wisely.

Coalition building

Individual teachers as well as small groups can have a positive impact. Marine education can be institutionalized across cultures, disciplines and organizations, so that its influence goes beyond a fortunate few classrooms. If the perspective of marine studies is broadened to include the importance of water in other subjects besides science, more comprehensive and sustained coverage will be possible. There also will be more groups who identify themselves as stakeholders and who may become active in promoting aquatic education. Professional organizations with an aquatic orientation (such as marine trades associations) should be urged to consider an active role. This might include preparation and distribution of materials, presentations at teachers' professional meetings or teacher and student awards. In teacher organizations, an "Inland Aquatic Education" section may be appropriate.

Funding

In times of uncertain financing for education at all levels, non-traditional or interdisciplinary courses and materials do not compete favorably with the basic disciplines. The best insurance for having some level of aquatic education in the curriculum is to infuse it into existing courses and established guidelines. Another approach is to develop creative funding arrangements, diversify support through foundations and agencies, develop new ways to use traditional funding sources, and build coalitions with education interests outside the classroom (Crane, et al. 1988). It may be helpful for an appropriate teacher organization to sponsor a workshop on funding sources and alternatives and to produce a publication on strategies and models for securing funding.

Publicity & promotion

It is an understatement that marine education efforts lack visibility in most regions not situated directly on the water.

"Where have you been all my life?" frequently is the response when teachers find out about NMEA, Sea Grant programs, or other marine education activities. It is an understatement that marine education efforts lack visibility in most regions not situated directly on the water. Aquatic education needs to be seen as an interdisciplinary concern that satisfies curriculum guidelines and is an important part of K-12 coursework. Some suggested means of recognition:

1) States or districts should declare a "Marine Education Week," "River Education Week," or whatever is appropriate to the local effort. Coordinators in each district should encourage special programs, displays and activities that would involve the entire community. Existing Coast Week activities should encourage a parallel inland effort.

2) States and districts, professional organizations and/or state agencies should establish an Award for Teaching Excellence in aquatic education. Announcement of winners could coincide with the Marine Education Week proposed above.

3) In each district, educators should develop a brief paper demonstrating how marine education materials can be integrated into various curricula. This paper should be sent to appropriate administrators. It should characterize marine education as an integral part of the curriculum rather than an expendable special interest topic.

Perhaps the collective wisdom of the educators, administrators, agency representatives, businesses and government representatives in the Great Lakes will be of value for inland aquatic education wherever it emerges. Enthusiasm and determination can help an individual teacher or group overcome many obstacles.

References

Goodwin, H.L. & Schaadt, J.G. (1978). *A statement on the need for marine and aquatic education.* Newark, DE: University of Delaware.

Fortner, R.W. (1980). You have to have an ocean... *Science and Children, 18*(2), 38-40.

Lanier, J. (1988). The national marine educators' association historical notes. *Current: The Journal of Marine Education, 8*(2), 5-19.

Disinger, J.F. (1987). *Environmental education in K-12 curricula (Information Bulletin No. 2).* Columbus, OH: ERIC Clearinghouse for Science, Mathematics and Environmental Education.

Western Regional Environmental Education Council. (1986). Project Wild - Aquatic.

Regan, A. (1982). *Directory of marine education resources.* Washington, DC: The Center for Environmental Education.

Sea Grant Marine Education Center. *MEMS, the marine education materials system.* Gloucester Point, VA: Author.

National Marine Educators' Association. (1988). Informational brochure. Kure Beach, NC: Author.

Crane, T. & Great Lakes Education Task Force. (1988). *Opportunities for expanding Great Lakes education in the region's classrooms. Final report of the Great Lakes Education Task Force to the Great Lakes Commission.* Ann Arbor, MI: Great Lakes Commission.

Seager, M.L. (1988). *Priorities of and constraints on teaching certain water topics in Ohio grades 5 and 9.* Unpublished master's thesis. Columbus, OH: Ohio State University.

Marine education: A bibliography of educational materials available from the nation's sea grant college programs. (1988). College Station, TX: Texas A & M Sea Grant College Program.

Cole-Misch, S. (1987). *Directory of Great Lakes education materials.* Windsor, Ontario: International Joint Commission.

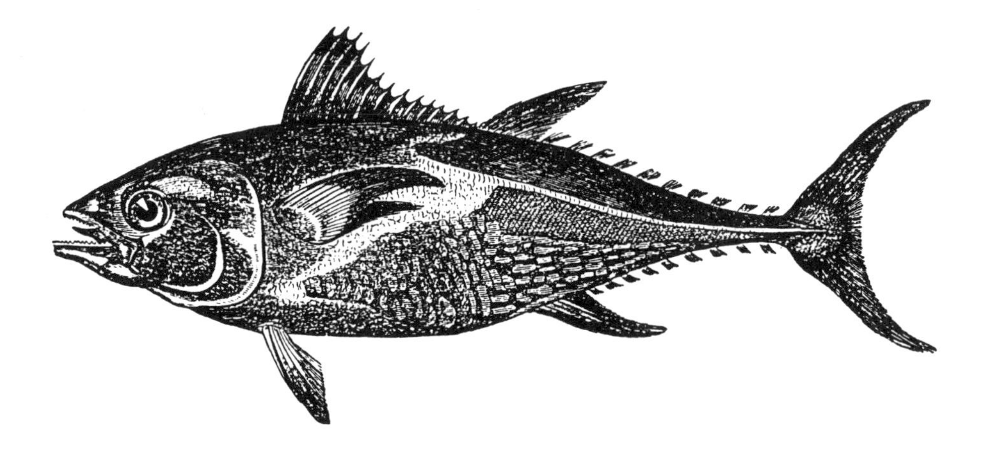

Investigating Inland Marine Heritage: Scientific & Technological Aspects

Norman D. Anderson

When asked to describe how inland marine biology teachers might have students investigate the scientific and technological aspects of marine heritage, my first thought was to describe investigations that could help them appreciate the heritage of the coastal states. But this is not where these students live and these teachers teach. They live in states like Kentucky, Wyoming and my native state of Iowa. Why not look at investigations that examine the role of science and technology in the marine heritage of inland states? For purposes of example, fewer states are any more inland than Iowa.

Investigations in marine biology are usually done in a laboratory or in the field. But there is another type that students may find appealing. It is the investigation of the marine history of a region, with special attention given to how people's actions have influenced the aquatic life forms found there.

Research Questions

Those who live in seacoast towns, many of which are more than 300 years old, may have difficulty comprehending how recently much of the inland section of our country was settled. What role did water play? First, there were no major rivers for transportation, so major settlement came after the railroad was built. Second, the glaciated lakes of Minnesota spilled over into parts of northwest Iowa and much of the area was covered by shallow lakes and sloughs. Only after the land had been drained by installing underground tile and open drainage ditches could farmers cultivate the land. With the spread of farming came towns, a few miles apart and within reach by horseback, buggy or wagon. What effect did draining the land have on the region's ecology? On spawning by various species? How were fish populations affected by dams and screens placed across streams draining the lakes? Old timers tell of catching enough fish in a day to keep a family supplied for weeks. How much truth is there to these stories, and if true, why is this no longer the case?

Ice fishing, as I recall, was illegal on most Iowa lakes during the 1930s and 1940s. This was a period of closed seasons and catch limits on most species, such as perch and walleye pike. How have fish management practices changed over the years? What advances in scientific knowledge produced these changes?

In 1909 the University of Iowa established the Iowa Lakeside Laboratory on the west side of Lake Obokoji not far from where I grew up. What kind of research has been done at Iowa's Lakeside Lab or similar labs such as Ohio State's Stone Lab at Put-in-Bay in Lake Erie or the University of Michigan Biological Station on Douglas Lake near Pellston? Although these labs are not as well known as those at Woods Hole in Massachusetts or Scripps in California, students will find them especially interesting because of their proximity. Investigating the contributions of these inland labs can be a worthwhile project.

My first teaching position was in the river town of Burlington in southeastern Iowa. Having grown up on a farm, I was particularly interested in the students whose families earned their living by commercial fishing on the Mississippi. How has this livelihood changed over the years? How did dredging the river and building dams and locks affect commercial fishing? What have been the effects of raising catfish and other species in ponds using methods based on scientific research?

Up the river a few miles from Burlington is the town of Muscatine, famous for its buttons made from clam shells. How did this industry get started, and how has it changed over the years?

I hope the above examples illustrate the possibilities for investigating marine heritage through the local history of inland regions. Teachers and students should have little difficulty identifying a list of similar questions about the marine heritage for their community.

Additional Investigations

Percy Bridgman, Nobel Laureate in Physics, has described the methodology of scientific investigation as "doing one's damnedest with one's mind, no holds barred." This also applies to the investigation of marine heritage, regardless of whether it be coastal or inland.

Perhaps the best known student investigations of local heritage in general are those done as part of the Foxfire Project in Georgia. Students at Cape Hatteras School on North Carolina's Outer Banks used the Foxfire approach in the 1970s to produce *Sea Chest*, a publication packed full of interesting accounts of their local marine heritage. These and other similar projects primarily used oral history; students can develop valuable skills through interviewing local residents and preparing written reports of their findings. They also may find it useful to review the files of local libraries, newspapers and historical societies.

Investigating any aspect of marine biology involves library research, whether it be by using *Biological Abstracts* to find reports of laboratory or field investigations or other reference tools to find material on marine heritage. Using the *Reader's Guide to Periodical Literature*, I found John Madson's intriguing article "Mississippi Shell Game" in the March 1985 issue of *Audubon*. The study of the "Great Pearl Rush" and how it turned an Iowa town into the capital of the button industry made me want to learn more.

People and organizations are another source of valuable information when doing research, regardless of the type. In doing research for this article, I called educators associated with Sea Grant Programs in North Carolina, Ohio and Virginia. They, in turn, suggested people I should contact in Michigan, Wisconsin and several other states. There are people doing research who are knowledgeable about almost every aspect of any topic, including inland marine biology. Making contact with these people is called "getting into the network." Encourage your students to write letters or, if money is available, make phone calls. Busy people will often give information over the phone that they don't have time to send in writing. It helps if students have done some preliminary work before they write or call. They should avoid letters like those sometimes received at science fair time: "Please send me everything you have on the moon, and I need it by next Friday."

There are people doing research who are knowledgeable about almost every aspect of any topic, including inland marine biology.

And, finally, don't overlook what may seem to be rather unconventional sources of information. Take the hobby of collecting old picture postcards or stereoview cards. These pieces of paper memorabilia often are the best surviving images of a local community's history, including its marine heritage. You can have fun using these collectibles to piece together a picture of the marine heritage of some of the places in your community.

These are but a few ideas for topics and methods for investigating the scientific and technological aspects of a local community's marine heritage. These suggestions hopefully will "prime the pump," and some of your students will get "hooked" on this type of research.

References

Anderson, N.D. (1984). Using picture postcards and other paper collectibles in marine education. *Current, The Journal of Marine Education, 6*, 6-9.

Madson, J. (1985). Mississippi shell game. *Audubon, 87*, 46-69.

Western Regional Environmental Education Council. (1987). *Aquatic, Project Wild (aquatic education activity guide)*. Project Wild, P.O. Box 10860, Boulder, CO 80308.

Wigginton, E. (1985). *Sometimes a shining moment: The foxfire section 1. experience.* New York: Anchor Press/Doubleday.

Supermarket Marine Biology

(Or: how to obtain marine realia)

E. Barbara Klemm

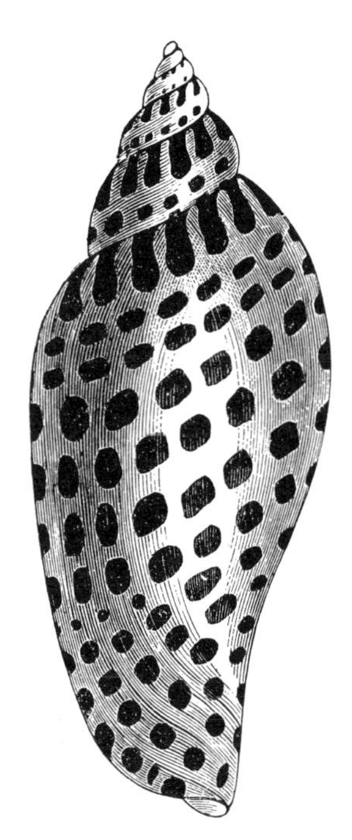

Teach marine science in an inland school? Why not? Effective teaching depends on what happens in the classroom—not on where the classroom is located. From research we know what works in science education: effective science teaching engages students in experimentation, allowing them to discover science concepts for themselves (U.S. Department of Education 1987).

Fifteen years ago the Hawaii Marine Science Studies (HMSS) program set out to develop a one-year, multi-disciplinary science course in which secondary students would investigate the marine environment (Klemm 1989, 1990). HMSS students use the same tools as professionals who study the marine environment and understand it best–the fundamental concepts and inquiry methods of biological, chemical, physical and geological oceanographers and related marine technologies. Content is taught via inquiry sequences centered around observation and experimentation with marine and aquatic animals, plants, water samples, sediments and other realia (concrete examples that can be touched and observed). The HMSS program is committed to the notion that investigations of tangible phenomena provide a basis for generalized understanding of the marine environment.

A key to successful marine education, then, is the availability of marine realia for the classroom. Input from more than 500 teachers in 26 states has convinced HMSS developers that marine science can be taught effectively by bringing authentic examples of the marine environment into the classroom. This chapter suggests ways science teachers can obtain marine realia.

Available Resources

Out of concern for conservation, teachers should restrict the use of living organisms to behavioral biology investigations involving organisms that are abundant, safely handled and hardy. A guppy, for example, may seem commonplace and quite a long way from a shark. But sharks' mating behaviors, feeding frenzies, live births and cannibalistic behaviors cannot be experimented within most classroom environments. Guppies' behaviors can. (Then give the students a shark tooth or a shark jaw to examine, and see what happens!)

Among the living organisms that teachers commonly collect from freshwater, brackish water, or saltwater include crustacea (crayfish, crabs, shrimp, lobsters) and mollusks (clams, mussels, snails). From the standpoint of understanding biology, it matters little which particular organism students observe. The organism serves as a specific, tangible entry-point for larger understanding of aquatic phenomena.

From the Supermarket

Many marine organisms are readily available because they are foods. These include a variety of freshwater and saltwater fishes; clams, oysters and mussels; shrimp, crabs and lobsters; squid and octopus; and, sometimes, seaweeds. Table 1 lists specimens you can find in fish markets, seafood restaurants, supermarkets and bait shops. Check gourmet shops and ethnic food stores for such "incredible edibles" as canned jellyfish or sea cucumbers and dried squid or seaweeds. Some marine education activities can involve students in cooking and eating.

Many other organisms or fragments of organisms can also be found in shops. Dried starfish, seashells and corals are often found in hobby stores, gift shops and department stores. Two types of natural sponges are commonly sold, one as a cosmetic sponge, the other as a cleaning sponge. Teachers can find sea urchin spines used in wind chimes and their tests (internal skeletons) used for night-lights and Christmas tree ornaments.

Visit stores to find out how they stock potential resources–alive, fresh, frozen, dried, or canned. Explain to proprietors that you teach marine science, and you may find them very helpful in obtaining not only locally-available organisms, but those shipped from other parts of the country.

From Ancient Oceans

Look also for relics from the ancient seas that covered much of the U.S.—fossilized shark's teeth and fish skeletons; crinoids; and coelenterates (coral), trilobites, sea urchins and bryozoans, to name a few. You may be able to collect some of these locally; otherwise, stores sell them. Some of the fossils are so well preserved they can be used like modern, fresh organisms in studying specific types of marine organisms. Fossils add the dimension of change over time to environmental inquiry.

From Sediments

In one of the most popular HMSS activities, students collect sands and sediments from all over the world, then use a stereo microscope to analyze specimens for their biological and abiological components. Do not overlook the microscopic radiolarians, diatoms and foraminifera that abound in muds, both modern and fossilized. Make students, parents, teachers and friends aware that you are collecting sands, and they will be brought to you–in film cans, envelopes and all sorts of containers. Be sure the sands are dried and labeled.

From Trading with Others

Set up a trading, lending or pen-pal system with teachers and students living elsewhere. Ask if notices can be placed in newsletters such as NABT's *News & Views*. Remember–what's commonplace in your area may be exotic elsewhere. Dried specimens such as seashells, sand dollars, crab molts, egg cases, sponges, coral and pressed seaweeds are typical of the kinds of organisms obtained through trading. Personal collections of rocks, sands, corals, seashells or fossils occasionally are donated.

Keep in mind that the value of traded or donated specimens is greatly enhanced if they are adequately labeled. In the process of trading, students can research and write information about local specimens and compare local examples with specimens from other locations. Plan to keep maps on hand, because students become interested in locating where specimens came from, thus leading into other marine lessons.

Collect Responsibly

Collecting living specimens for use in marine science classes ought not be pursued unless teachers and students are prepared to care for the organisms in classroom aquaria. The collection process ought to be a model of sound conservation practices and environmental ethics. Collect as few specimens as possible and

Table 1. Examples of how marine animals can be obtained from local stores.

Phylum	Specimen	Condition	Where Purchased
Porifera	Demospongae–cleaning sponge, facial sponge	dried skeleton dried skeleton	hardware, grocery store drug store
Coelenterates	corals jellyfish	dried skeleton (canned)	dept. store, gift shop, discount store (gourmet food shop)*
Worms	Annelids	live	bait shop
Echinoderms	starfish, sea urchin	(dried whole) (dried skeleton)	(toy dept., gift shop) (gift shop)
Mollusks	Bivalves Gastropods–escargot squid octopus	fresh, frozen canned frozen, whole (fresh, whole)	fish mrkt., restaurant gourmet shop, restaurant grocery store, fish mrkt. (fish mrkt.)
Arthropods	crabs, shrimp, lobster	fresh, frozen	grocery store, fish mrkt. (least $$$ at bait stores)
Fish	local fish	frozen fresh, whole live	grocery store, fish mrkt. fish mrkt. pet store

* Parentheses are used where specimens may be harder to locate.

only from abundant organisms. Observe fishing laws and other restrictions. Obtain permits when necessary. Once inquiry is finished, if organisms are not going to be eaten, raised, or further observed, return them to their natural environment.

Whenever possible, collect like a scavenger rather than a predator. Avoid killing living organisms for the purpose of adding to the class collection. Search for beached seashells, coral and seaweeds, and look for abandoned specimens like molts and egg cases. Use unwanted specimens–both dead and alive–that became hooked or trapped in nets and which fisherman normally would discard. Save the scales, skeletons, teeth and jaws from cleaned or dissected fish. Collect shucked clam, oyster, mussel and abalone shells. Keep in mind that imperfect, broken and partial specimens still have instructional value.

Learn how to preserve and store specimens so they can be used over a number of years. Fresh specimens may be stored in a freezer, but not indefinitely. Some specimens, like sponges, sea stars or sea urchins, can be slowly dried in ovens set at low temperatures. Other specimens, like fish, must be preserved chemically. Seaweeds may be pressed and then arranged artistically or in formal botanical collections.

Find "buddy" teachers willing to join you in collecting organisms. Share your collections. For example, one teacher could maintain a sand collection and another a mollusk collection. Some HMSS teachers arrange trades of entire sets of prepared instructional materials. One teacher chooses a topic such as coral and organizes the specimens and prepares handouts, overhead projections and bulletin board displays. After teaching the unit, the entire set of materials is swapped with another teacher who has prepared a set of instructional materials on another topic. HMSS teachers who work cooperatively as buddy teachers say they enjoy exchanging not only materials but also ideas with colleagues. Trading sets of realia saves them time and conserves organisms.

HMSS teachers who work cooperatively as buddy teachers say they enjoy exchanging not only materials but also ideas with colleagues.

Collecting marine realia is addictive and contagious. Just let it be known that you want marine specimens for your class and students, parents, other teachers and friends will join in. They will bring you specimens from all over the world and be curious about what they've found. Obtaining marine realia is possible no matter where you teach. Try it!

References

Klemm, E.B., Pottenger, F.M., Reed, S.A. & Coopersmith, A. (1990). *The fluid earth: Physical science and technology of the marine environment* (3rd. ed.). Honolulu, HI: HMSS Project, Curriculum Research and Development Group, College of Education, University of Hawaii.

Klemm, E.B. (Ed.). (1982). *The living ocean: Biological science and technology of the marine environment* (2nd ed.). Honolulu, HI: HMSS Project, Curriculum Research and Development Group, College of Education, University of Hawaii.

U.S. Department of Education. (1987). *What works. Research about teaching and learning.* Washington, DC: U.S. Government Printing Office.

Setting Up a Marine Aquarium

Nancy Chartier Balcom

A marine aquarium can open up a whole new world of exploration to students. Studies of the biology and physiology of invertebrates and fish can be integrated with studies of the chemical and physical properties of the tank water. Students can also be encouraged to take responsibility for the maintenance of the tank and the care of the animals. A marine aquarium is a more complex system to maintain than a freshwater aquarium. This chapter can be used as a step-by-step guide to setting up a saltwater aquarium in your classroom. There are suggestions of ways to explore the aquarium at the end of this chapter.

Setting Up: Equipment

The tank

The tank should hold at least 20 gallons of water to provide for good water circulation and bacterial action and to prevent animal deaths due to overcrowding. Before you buy a tank, check newspaper ads for used aquaria equipment. If feasible, check with local natural history museums or aquariums to see if any have a "rent-a-tank" program. Ask around; perhaps someone at school or in the community could loan you a tank temporarily. It should be all glass or have a plastic frame, as metal corrodes. It should also be "low and wide" rather than "high and deep" so that surface area is maximized. The tank should be well-supported (seawater weighs 8.5 pounds/gallon) and set up away from sunlight, drafts and heat sources. Check for leaks.

Subgravel filter

The filter is a perforated plastic plate which completely covers the bottom of the tank. It supports the gravel while allowing water to circulate under the plate. This circulation prevents the formation of stagnant areas caused by the build-up of hydrogen sulfide, methane and carbon dioxide. The filter-plate should have holes in the back two corners for rigid tubing (airlifts), about an inch or so in diameter. These airlifts house the airstones. Crushed oyster shell, for the gravel bed, can be purchased at a feed store; it is a good calcareous substrate and inexpensive. Rinse the shell thoroughly to remove the dust before placing it in the tank; cover the filter plate with at least three inches of shell.

Air pump

An air pump is needed to put oxygen into the water. The pump should be set above the tank to prevent backflow. Connect the pump to a gang valve, using 3/16-inch diameter flexible polyethylene airline tubing. Connect flexible tubing from two of the valves to rigid airline tubing (about same diameter), cut approximately two inches shorter than the airlifts. Use flexible airline tubing to connect the airstones to the thin rigid tubing and place one airstone in each airlift. The valves should be adjusted once water is in the tank so that both airstones briskly produce minute bubbles. This creates the water circulation in the tank.

Outside filter

A plastic filter containing activated carbon that hangs on the outside of the tank is an optional item but is very useful in maintaining the health of a marine aquarium.

Heater

The tank water should be kept at a constant temperature; it may be necessary to place an immersion heater in the tank. Read the manufacturer's instructions for placement and setting heater temperature.

Salt water

You will need to make artificial seawater. One brand of sea salts commonly used is Instant Ocean. Fill a large container (not the tank) with fresh water. Place an airstone in the container and let the fresh water bubble for 24 hours to eliminate any chlorine in the water. Add the required amount of sea salts, mix and aerate until the water turns clear. Then carefully add the water to the aquarium. The amount of salt used depends on the salinity required for the animals you intend to keep.

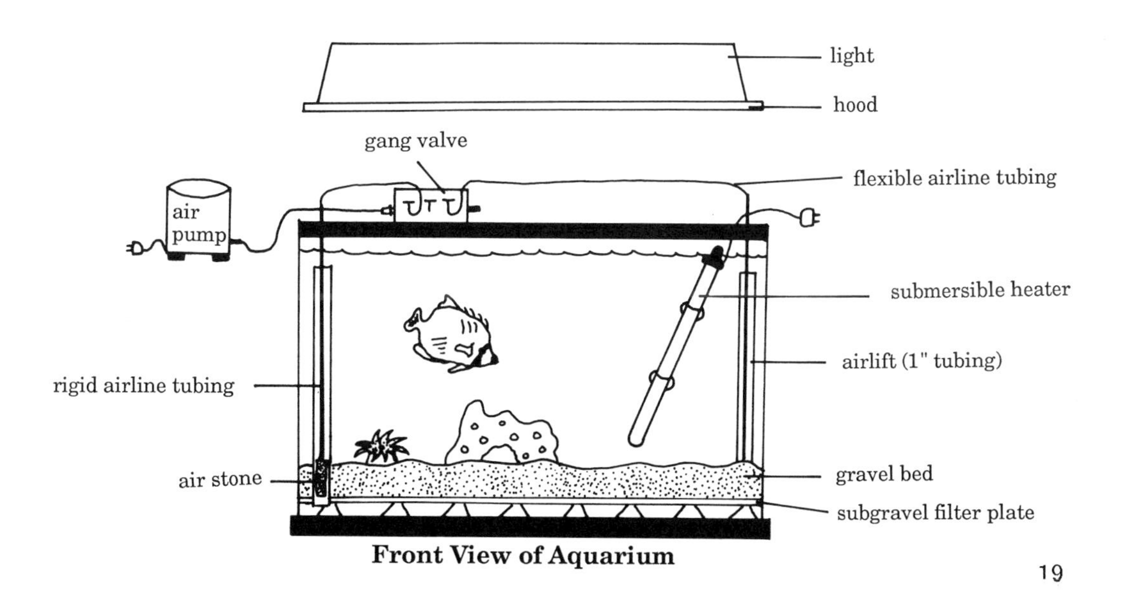

Front View of Aquarium

Hood

A plastic hood (commercial tank hood, clear plexiglass sheet) reduces evaporation, keeps animals from jumping and provides an added safety feature. To light the tank, use a commercial aquarium light since it is completely enclosed to prevent salt spray corrosion. Lights help show fishes' true colors.

Other hints

Put rocks, shells, plastic plants, ceramic flower pots, etc. in the tank to give the animals a place to hide. Never move a tank that is even partially filled with water; it may start to leak.

Conditioning & Maintaining the Saltwater System

A word of caution

When working with the aquarium in a classroom, it would be prudent to unplug all equipment before sticking a net or your hand into the tank to avoid the possibility of electrical shocks. Salt water is extremely corrosive and is a good conductor. Make sure, however, that everything gets turned back on. On that same note, avoid placing hands directly into the tank as body oils, soaps and lotions contaminate the water. If you wish to observe an animal more closely, place it in a smaller container of water for a short period, then return the animal to the tank and discard the water.

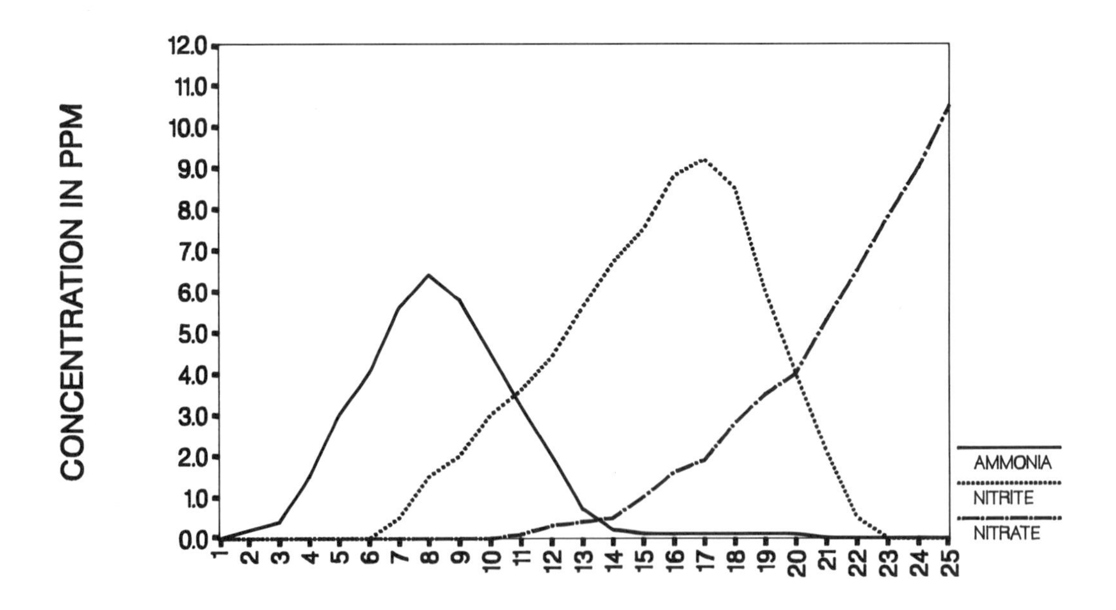

Figure 1. As the tank conditions, the ammonia concentration peaks first, then the nitrite and finally the nitrate.

Conditioning

Once the tank is set up with salt water, the water must be "conditioned" before animals are added. A biological filtration system must be established, which requires the presence of certain bacteria. One way to condition a tank is to put a cupful of gravel from an already established marine aquarium in the tank and let it run for a week or two. Or, you could put some hardy (insensitive to ammonia) fish or invertebrates in the tank and feed them for a period of a week or two. Decaying food and animal wastes support the production of bacteria. Do not rush the conditioning process or the result will be the loss of your animals.

Biological filtration refers to the nitrogen cycle. Bacteria break down animal waste products (ammonia) into nitrite and nitrate. Ammonia and nitrite are toxic to most marine animals; nitrate is less toxic at low concentrations. As a tank "conditions," the ammonia concentration peaks first, then the nitrite and finally the nitrate (Figure 1). Test kits, available at most aquarium supply stores, can be used to monitor the changes in the nitrogen cycle in your aquarium. The tank is safe for animals when the ammonia concentration is less than 0.1 ppm and the nitrite concentration has peaked.

Adding animals

What you put in your aquarium depends on your location and what is available. If you live near the coast, you have more options than inland teachers when looking for animal sources. I do not recommend that beginning aquarists try to maintain the brightly colored tropical species that many pet stores sell–they are expensive and difficult to care for. Look for hardier saltwater species of fish or invertebrates such as silversides, mummichugs, killifish, sea bass, spider crabs, snails, starfish and hermit crabs. (See resources listed at the end for some mail order possibilities.) Check with local aquariums for suggestions and perhaps sources, ask at reputable pet stores, or ask community home aquarists for assistance and help.

Avoid overcrowding the tank. One rule of thumb is three inches of fish for every square foot of tank surface area. A few healthy animals are preferable to many dead ones. Animals should be acclimated to a new tank. If they have been transported in plastic bags, untie the bag, roll the top down and float the bag in the aquarium. Every 15 minutes or so, dip the bag into the aquarium to add water. After at least an hour of acclimation, tip the bag and let the animal swim into the tank. If an animal looks stressed while in the plastic bag, drop an airstone into the bag and aerate gently.

Feeding

Once you have put animals in the tank, you need to feed them. Overfeeding is a primary cause of tank fouling, so every other day only put in as much feed as the animals can consume in five to ten minutes. Use a net to scoop out uneaten food. The food you use depends on the animals; again, check with reputable pet stores. Keep records of feeding times, foods and whether each animal ate.

Water tests

The tank water should be tested weekly to ensure that its condition is healthy. Make up a record sheet and delegate this responsibility to your students.

1. pH–acceptable range: 7.5 to 8.3.
2. Temperature–should be kept constant, within the proper range for the animals kept.
3. Specific gravity–should not vary more than ±0.002 from the correct value (Table 1). Use a hydrometer to measure specific gravity: turn off the air to the aquarium, carefully lower the hydrometer into the water and take a reading at the meniscus when the instrument stops bobbing.
4. Ammonia–acceptable range: less than 0.01 ppm.
5. Nitrite–acceptable range: less than 0.1 ppm.
6. Nitrate–acceptable range: less than 20.0 ppm.
7. Dissolved oxygen–acceptable range: 5.0 ppm is the absolute lowest limit.
8. Water changes–change 10 to 20 percent of the aquarium water approximately every two weeks. If the water looks exceptionally cloudy or the pH, ammonia, nitrite, or nitrate

Table 1. Specific gravity of seawater for selected temperatures and salinities.

°F	°C	34 ppt	32 ppt	30 ppt	28 ppt	25 ppt	32 ppt
85	30	1.021	1.020	1.018	1.017	1.014	1.011
84	29	1.021	1.020	1.018	1.017	1.015	1.011
82	28	1.022	1.020	1.019	1.017	1.015	1.011
81	27	1.022	1.020	1.019	1.017	1.015	1.011
79	26	1.022	1.021	1.019	1.018	1.016	1.012
77	25	1.023	1.021	1.020	1.018	1.016	1.012
75	24	1.023	1.021	1.020	1.018	1.016	1.012
73	23	1.023	1.022	1.020	1.019	1.016	1.013
72	22	1.023	1.022	1.020	1.019	1.017	1.013
70	21	1.024	1.022	1.021	1.019	1.017	1.013
68	20	1.024	1.023	1.021	1.020	1.017	1.014
66	19	1.024	1.023	1.021	1.020	1.018	1.014
64	18	1.025	1.023	1.022	1.020	1.018	1.014
63	17	1.025	1.023	1.022	1.020	1.018	1.014
61	16	1.025	1.024	1.022	1.021	1.018	1.014
59	15	1.025	1.024	1.022	1.021	1.018	1.015
57	14	1.025	1.024	1.022	1.021	1.019	1.015
55	13	1.026	1.024	1.023	1.021	1.019	1.015
54	12	1.026	1.024	1.023	1.021	1.019	1.015

concentrations are high, change 25 to 50 percent of the water (depending on severity). Scrub the walls of the tank to remove algae. Replace the activated carbon in the outside filter every three weeks or so.

Investigating a Marine Aquarium

With care and some luck, you should be able to maintain a marine aquarium in your classroom. If you run into problems, consult a pet store employee, local aquarium staff, or home aquarist. A good basic reference is listed at the end of the chapter (Spotte 1973). It is a useful beginner's guide to setting up and maintaining an aquarium. Good luck and enjoy!

Activities

Here are some suggestions of activities to explore and develop that focus on a marine aquarium.

Chemical

1. Measure changes in the concentrations of ammonia, nitrite and nitrate in the aquarium from set-up through the addition of animals. Plot changes in the concentrations over time. Monitor the concentrations weekly after animals are added to make sure the conditions stay within the acceptable range for the animals. Compare your results with Figure 1.
2. What is the difference between salinity and specific gravity?
3. Measure the pH of the tank water weekly. How does an overly heavy feeding or an animal death affect the pH of the tank water? How does sodium bicarbonate (baking soda) affect the pH of salt water?
4. Measure the dissolved oxygen (DO) concentration in the water. Boil some fresh water and some salt water. What happens to the dissolved oxygen concentration? How does water temperature affect dissolved oxygen: does cold or hot water hold more DO? What would happen to the aquarium water if too much oxygen were pumped into the water (i.e., discuss saturation, supersaturation, etc.)?

Physical

1. What is the turnover rate of the water in the tank? Ideally, the total discharge from all the airlifts should equal the rate of 1 gallon/sq. foot surface area/minute. Determine the surface area and volume of the tank. Hold a measuring cup under one airlift and time how long it takes for four cups to be filled. (Four cups equal 0.25 gallons.) Calculate how much water is filtered in one minute. What is the gpm/sq. ft.? For example, if the volume is 2.5 sq. ft. and four cups are filled in 10 seconds, then 1.5 gallons are filtered in one minute. The gpm/sq. ft. is 0.6. What about the other airlift? [Copyright ©1973 John Wiley & Sons, Inc. Reprinted by permission of John Wiley & Sons, Inc. from Spotte, S. (1973) *Marine aquarium keeping: The science, animals, and art.*]

2. Using extra sea salts, make up several different concentrations and use food coloring to dye them different colors. Watch the effect of carefully adding different concentrations of salt water to fresh water. Try this with different temperatures of water. Discuss buoyancy and currents.

Biological

1. Discuss the nitrogen cycle.
2. Discuss osmoregulation. Do saltwater fish drink water? Do freshwater fish drink water?
3. Depending on the animals kept, explore different phyla, feeding habits, social interactions, predator/prey relationships, coloration and camouflage, competition, adaptations, habitats and food chains.

Sources for Marine Animals

Northeast Marine Environmental Institute (NEMEI), Basic Marine Life Collections, P.O. Box 666, Monument Beach, MA 02553.

Carolina Biological Supply Company, 2700 York Rd., Burlington, NC 27215

References

Allen, W. & McLaughlin, P. (1985). *Sea sampler: Aquatic activities for the field and classroom (secondary) (South Carolina Sea Grant Publication SC-SG-TR87-2).* USC Baruch Marine Laboratory, P.O. Box 1630, Georgetown, SC 29442.

Barker, W. (1981). *A guide to field studies for the coastal environment (grade 8) (Project CAPE Publication SC2).* Dare County Board of Education, P.O. Box 640, Manteo, NC 27954.

Butzow, J. (1979). *Northern New England marine education project, K-12 multidisciplinary units.* Orono, ME: University of Maine at Orono, College of Education and Maine-New Hampshire Sea Grant.

Coon, H. & Price, C. (1977). *Water-related teaching activities, K-12.* Columbus, OH: ERIC Center for Science, Mathematics, and Environmental Education, Ohio State University.

Florida Oceanographic Society. (1980). *The source book of marine science.* Stuart, FL: Author.

Greene, D. *Aquatic activities for youth.* Ithaca, NY: New York Sea Grant Cooperative Extension and Cornell University. Write: D. Greene, Youth Education, 21 Grove St., East Aurora, NY 14052.

Hastie, B. (1981). *Water, water everywhere... marine education in Oregon. A guide to instruction about fresh and salt water.* Salem, OR: Oregon Department of Education, 700 Pringle Parkway SE, Salem, OR 97310.

Hunt, J. (1980). *Marine organisms in science teaching.* College Station, TX: Texas A & M University, Sea Grant College Program.

Mauldin, L. & Frankenberg, D. (1978). *North Carolina marine education manual (Vol. 1-5).* Raleigh, NC: North Carolina State University, UNC Sea Grant.

National Sea Grant College Program. (1988, July). *Marine education: A bibliography of educational materials available from the Nation's Sea Grant College Programs.* (TAMU-SG-88-401-R). College Station, TX: Texas A&M University.

Spotte, S. (1973). *Marine aquarium keeping: The science, animals, and art.* New York: John Wiley & Sons.

Section II. Activities: Marine Biology

Using Marine Aquaria

David A. Wright
Emmett L. Wright

A marine aquarium in the classroom brings a bit of the living ocean to students and provides a potential focus for the adolescent's boundless fascination and interest in the sea. Using modern techniques, the establishment and maintenance of a marine aquarium is as easy as fresh water aquaria (Mowaka 1979). A marine aquarium is more expensive to set up, but well worth it. This chapter offers suggestions for using marine aquaria as instructional tools; books on setting up an aquarium are listed in the references and refer to the previous chapter.

Class Activities

A living marine system can be a valuable tool for discussion and learning. Instruction can begin before the aquarium tank is set up and operational. Discussions about water salinity, temperature, light and aeration could easily fit into a lesson on ocean chemistry. A lesson can also be taught on the types of equipment needed to maintain the aquarium and on its operation.

Once the aquarium is established, student interest can be directed by a required notebook of observations. This can consist of things the students observe on their own or specific changes in abiotic and biotic factors you want them to record over time. Most observations will fall into three major areas: water chemistry, observations of specific organisms and interactions among organisms (Spence & Medlicott 1982).

A properly established marine aquarium is, in a sense, a balanced living system. This balance depends on system inputs and outputs. As in the ocean, both abiotic and biotic factors will influence the aquarium's water chemistry. To successfully establish a marine aquarium you must have naturally occurring nitrifying bacteria in the filter media. These bacteria remove toxic nitrogenous wastes (ammonia and urea) by converting them to less toxic forms (nitrite then nitrate) (Mowaka 1981). Using water testing kits (i.e., HACH or LaMotte) students can measure the concentration of different nitrogenous wastes over time. Have students investigate if these concentrations change over time as the tank matures.

Algal growth is another indicator of the tank water chemistry. Have students use a microscope to observe and identify the progression of different types of algae blooms. Diatoms, growing along the bottom as a brown mat, will be first to colonize the tank, followed by blue-green algae. As the tank becomes established and the ammonia and nitrite levels decrease, these algae will be replaced by green algae (de Graaf 1973). Have students investigate this relationship between water quality and algal population speciation.

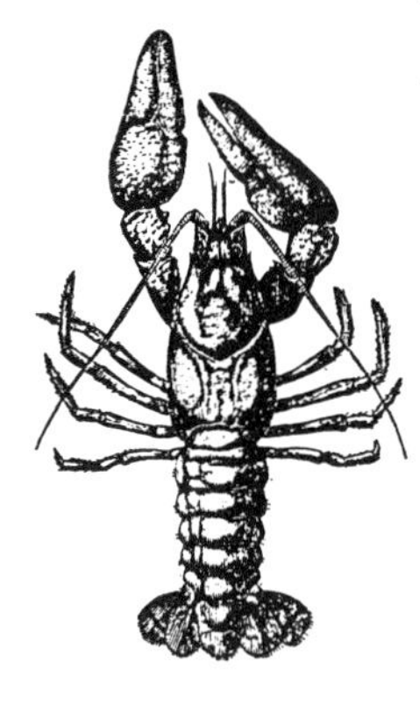

A major theme in biology is the adaptation of life forms through natural selection to fill various niches and accommodate changing environmental conditions. Marine animals, with their greatly varying forms and life styles, are excellent subjects for studying this aspect of biology. Have students examine animal body shape and coloration. Have them think about and answer: To what type of habitat and life style is this animal adapted? How do the features observed enhance its survival? What special problems does the marine animal face to survive?

Have students examine and compare different animals' individual features such as coloration (camouflage, countershading, disruptive, false eye spots, advertising and warning), body shape, appendages, special body features and behavior.

Design labs where students observe animals for set intervals over time. Have them record behaviors exhibited by a particular animal (i.e., eating, aggression, defense, motion, location in the tank). Some animals and their behaviors are best observed in individual containers such as large glass jars or plastic beverage containers. Be sure to return animals to the main tank when you are through so they will not die from fouled water.

Some animals are hardy enough to be used in manipulations (**Note:** be sure animals are handled gently and are immediately returned to the tank if they exhibit signs of stress). Hermit crabs can be trained to locate food in a simple T-shaped maze (Spence & Medlicott 1982). Fiddler crabs can be tested for food chemoreceptability. Place them out of water in the middle of a circular disk (8 inches) and place food at the edge of the disk at different angles of orientation to the crab (Hampton & Weston 1982). Test the pulling strength of a marine snail by gluing a string to the shell (using water-soluble glue) and attaching varying numbers of paper clips to it. Test animal preference for color by placing different colored pieces of paper under clear-bottomed observation trays. Crustaceans and some fish exhibit phototaxic responses (Spence & Medlicott 1982). Test their reaction to light.

Observe the novel way starfish move along the glass of the aquarium. How is this similar or different than the method used by sea urchins or other marine animals? Be careful. Moving the starfish when it is attached to the glass will injure the animal by ripping off some of its tube feet (Morholt & Brandwein 1986).

In a separate small tank, observe how clams pump water in and out of their bodies. Use an eye dropper to deliver a diluted (1 drop/liter) sample of food coloring in the water near one of the siphons. Which siphon draws water in and which directs water out of the animal (Hampton & Weston 1982)? Other bivalves (i.e., oysters) can be examined and a bit of india ink or soot used instead of food coloring (Morholt & Brandwein 1986). Observe the feeding behavior of other filter feeders (i.e., tubeworms, barnacles). How does it differ from bivalves?

Marine animals can be used to investigate interactive behavior between animals. Have students observe two animals' behavior over time. Which is aggressive or submissive? Do animals differ in their responses? Are any animal interactions mutually beneficial? Which in the relationship benefits most? Students can observe examples of interactive behavior. For example: fish (i.e., Damsels) may set up territories and defend them. One animal may kill and/or eat another (i.e., sea anemone and fish, starfish and shellfish). Crabs usually are aggressive and may dominate the tank. Some small fish (i.e., cleaner wrasse) clean parasites from larger fish. Clown fish or hermit crabs and sea anemones often form symbiotic relationships.

Sources of Marine Animals

Unlike coastal dwellers, landlocked science teachers cannot run to the seashore to obtain animals to stock an aquarium. Most large biological supply houses will ship live marine animals anywhere in the country. These usually come in sets and consist of the more hardy animals. Several animals representing different phyla often can be obtained in a single set. Many pet stores that trade in tropical fish can supply you with marine animals and allow you to observe the animals, select exactly what you want and ascertain their health before buying. Check grocery stores for live oysters, mussels, clams and lobsters. **Note:** lobsters are expensive and may require a refrigerated tank.

A marine aquarium in the classroom provides science teachers a wonderful resource and inland students firsthand experience with marine organisms. In addition, due to its diverse components, the marine aquarium can easily be incorporated into a wide range of topics for instruction.

References

Campbell, G. (1976). *Salt-water tropical fish in your home.* New York: Sterling Publishing.

de Graaf, F. (1973). *Marine aquarium guide.* Harrison, NJ: Pet Library LTD.

Hampton, C.H. & Weston, T. (1982). *Marine organisms in the classroom.* (Project CAPE [teaching module] SC1). Manteo, NC: Dare County Board of Education.

Hunt, J.D. (1980). *Marine organisms in science teaching* (TAMU-SG-80-403). College Station, TX: Texas A & M University Sea Grant Program.

Morholt, E. & Brandwein, P.F. (1986). *A sourcebook for the biological sciences.* San Diego: Harcourt Brace Jovanovich.

Mowaka, E. J., Jr. (1979). *An introduction to the marine aquarium.* (8141 Tyler Blvd.,) Mentor, OH: Aquarium Systems.

Mowaka, E.J., Jr. (1981). *The seawater manual: Fundamentals of water chemistry for marine aquarists.* (8141 Tyler Blvd.,) Mentor, OH: Aquarium Systems.

Spence, L. & Medlicott, J. (1982). North Carolina marine education manual. Raleigh, NC: North Carolina State University.

Marine Biology in Action–I

Observational laboratory experiences form the basis for Oceanography, a one-semester elective high school science course offered by the Benton County R-1 School District in Missouri. Students enrolled in Oceanography begin by preparing a standard 10-gallon aquarium for marine organisms.

The students establish a biological filtration system by introducing marine bacteria from another marine aquarium or by adding a small amount of ocean water. They also add an external filtration system to provide additional cleansing power and surface agitation and wash marine gravel to remove excess dust and silt. Students then establish the proper salinity conditions for marine organisms based on instructions included with a commercially available marine salt kit and establish proper temperature conditions for marine organisms. This preparation of the marine aquarium is completed by placing a full-cover lighted hood over the entire aquarium to reduce evaporation, increase environmental stability and improve viewing conditions.

Students keep a daily record of the preparation, establishment and set-up time, beginning shortly after the aquarium is first filled with water. This continues for the remainder of the course. Our students maintain an Observations Journal in which they record and graph the positions of all the organisms at a given instant; determine the salinity, temperature and pH values using a hydrometer, thermometer and litmus paper; and monitor the state of the NH_3/NH_4^+ system in the aquarium using a commercially available test kit.

Based on their observations, students then make comparisons and develop inferences involving the effects of environmental conditions, which may change quickly and dramatically in 10-gallon systems. Students also record any adjustments made to the system in order to maintain normal range values for the monitored parameters, as well as the behavior of the organisms before, during and after any adjustments have been made.

After several months, using the information contained in their journals, students conclude their laboratory experience by preparing written reports which describe and explain their observations, complete with charts and graphs. They also share their newly gained knowledge and the living organisms in the marine aquarium by presenting a "A Trip to the Ocean" program to elementary school students.

–John R. Sode

Section II. Activities: Marine Biology

The Anatomy of a Crab

William R. Hall, Jr.

Crabs are interesting creatures that range in size from several millimeters to a record four meters plus for the giant spider crab of Japan, *Macrocherira kaempferi*.

Life Cycle

Crabs are big business; a multi-million dollar business in the U.S. alone–king crab in Alaska, Dunginess in California, stone crab in Florida and Blue crab in the Mid-Atlantic states.

Like all arthropods, crabs have an exoskeleton composed of chitin which, in the case of crustaceans, is greatly strengthened by the deposition of calcium salts. While a suit of armor has its advantages, it also limits growth. Therefore, in order to grow, arthropods must molt or shed their exoskeleton.

The following lab exercise uses the Blue crab (one of the few animals that you can dissect that you also can be served in a restaurant). Blue crabs are familiar to many who live or visit Chesapeake Bay or other Atlantic estuaries. Approximately 50 percent of the value of the U.S. crab industry depends on the Blue crab catch. Its availability at seafood outlets is challenged only by the American lobster.

Molting begins in crabs with the splitting of a suture between the carapace (top shell) and the abdomen which is folded underneath. The crab then backs out of its old shell and pumps in water to stretch the new shell, which is soft and wrinkled. Crabs increase approximately 30 percent in size with each successive molt. Some species continue to molt for as long as they live with the length of time between molts increasing as the crab's size increases. Others molt a specified number of times before reaching sexual maturity with a terminal or final molt.

Fertilization is internal and occurs soon after the molt. The

The Blue Crab

Kingdom:	Animalia
Phylum:	Arthropoda (jointed foot)
Class:	Crustacea (hard shell)
Order:	Decapoda (ten feet)
Suborder:	Brachyura (short tail), "the true crabs"
Genus:	*Callinectes*, (beautiful swimmer)
Species:	*sapidus*, (savory)

fertilized egg then passes through the seminal receptacle to the outside and is attached to the pleopods by a sticky substance. Females in this state are often referred to as sponge crabs. Depending on the species and the water temperature, the eggs mature over a period of several weeks or longer, hatch and are released to the surrounding currents.

In the planktonic stage the larva look nothing like the adults and are called "zoea." As with all plankters, they drift at the mercy of the currents as they feed and grow. Finally they molt into a "megalopa," a half zoeal, half-crab stage and leave the plankton assuming the benthic life that we associate with crabs.

Dissection

Objectives

After this lab, students should be able to:

- Identify the crab's major external structures
- Identify the crab's major internal organs
- Describe how crabs respire, feed and reproduce
- Discuss the classification of the Blue crab

Materials

Blue crabs are available from local seafood retailers and are the ideal specimen since you can eat your lab. (Buy crabs already steamed or steam them yourself for 25 to 30 minutes. Make sure students wash their hands before handling crabs and have them use utensils from home when touching meat that will be eaten.) Preserved crabs are available from Carolina Biological Supply Company. However, since these cannot be eaten, it detracts from the experience.

Procedure

External anatomy–locate, identify and list the function of each of the following (see Figures 1-6 for diagrams; illustrations are adapted from Pyle & Cronin 1950):

carapace (cephalothorax)
cervical groove (marks boundary between head & thorax)
abdomen
telson
anus
maxillae 1 & 2
sternum (composed of sterna)
seminal receptacle (female)
stalked eyes
antennae
chelipeds (1 pair)
walking legs (3 pair)
swimming legs (1 pair)
maxillipeds 1, 2, 3
mouth
mandibles
antennules
pleopods (male, female)
penes (male)
inhalant opening (in front of chelipeds)
exhalent current (next to mouth)

Notes

You may wish to tape the external parts to a sheet of paper and label each one. The maxillipeds, thoracic appendages, are in reverse sequence. That is, maxilliped 3 is removed first, followed

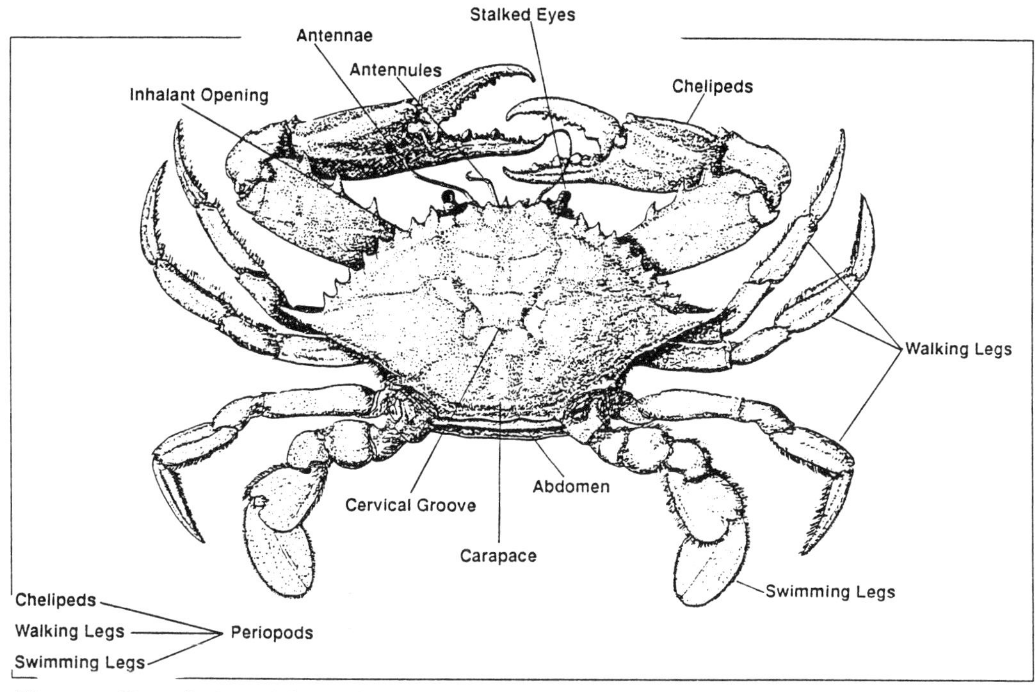

Figure 1. Dorsal view of the male of *Callinectes sapidus* Rathbun.

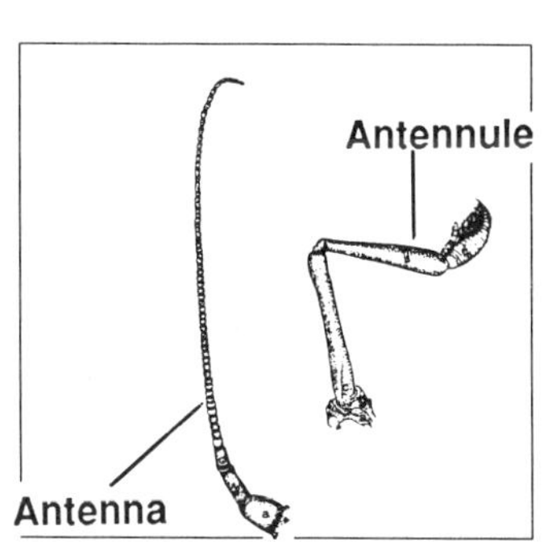

Figure 2. The sensory appendages. Antenna left, Antennule right.

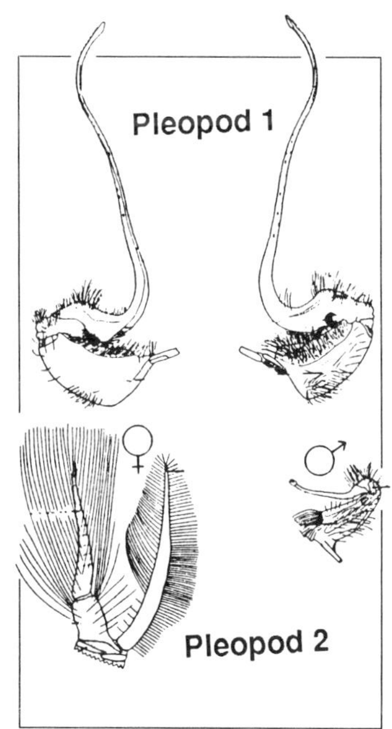

Figure 3. The first and second abdominal appendages. Lower left figure is female, others male.

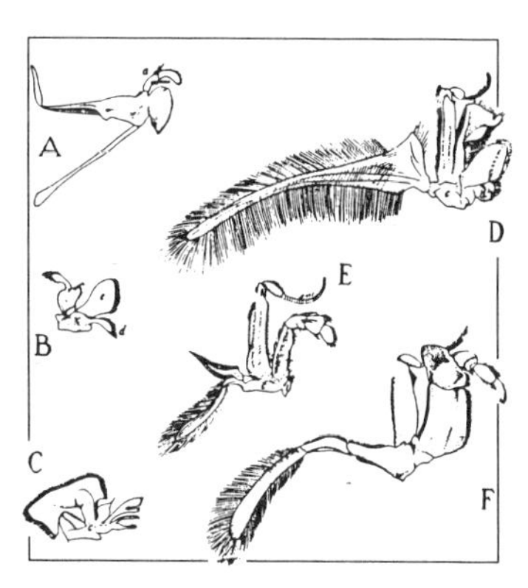

Figure 4. The mouth parts of *Callinectes sapidus* Rathbun: A. Mandible; B. First Maxilla; C. Second Maxilla; D. First Maxilliped; E. Second Maxilliped; F. Third Maxilliped.

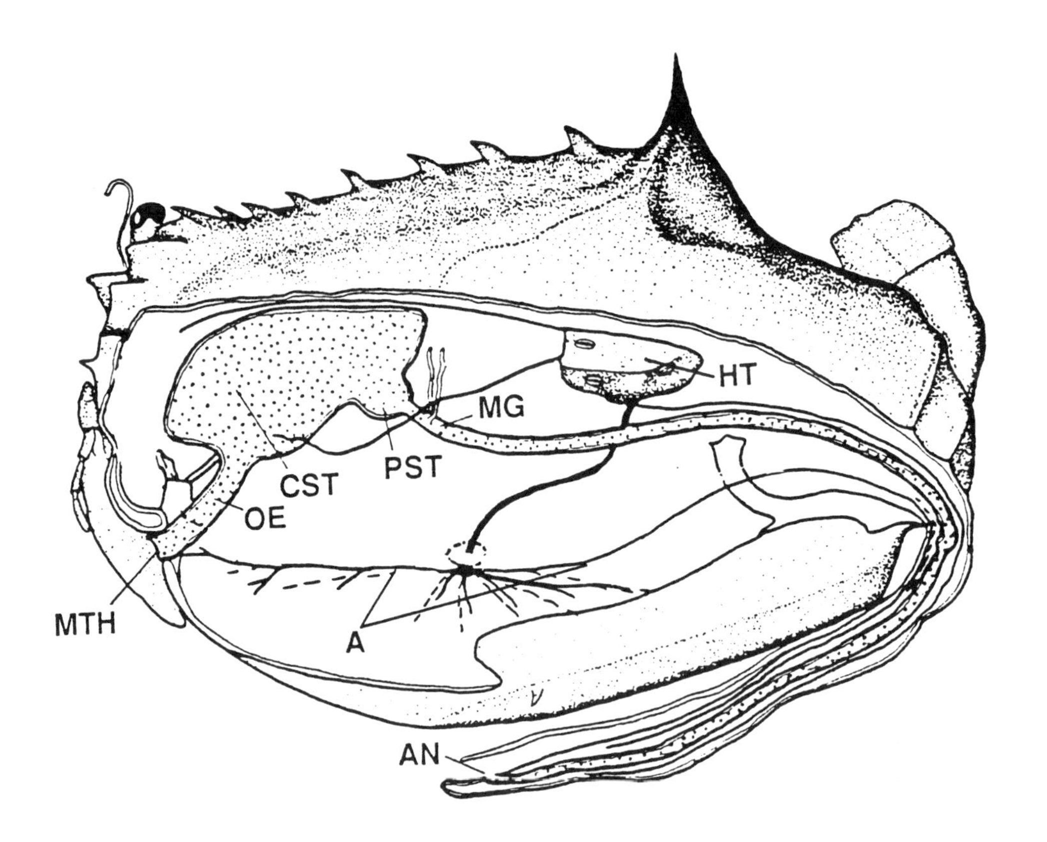

Figure 5. Diagram of a sagittal section through the crab from a dorsolateral aspect to show the relationships between the digestive and circulatory systems. The digestive system is stippled, the arteries solid lines. AN–Anus; CST–Cardiac stomach; A–Artery; MG–Midgut; MTH–Mouth; OE–Oesophagus; PST–Pyloric stomach; HT–Heart.

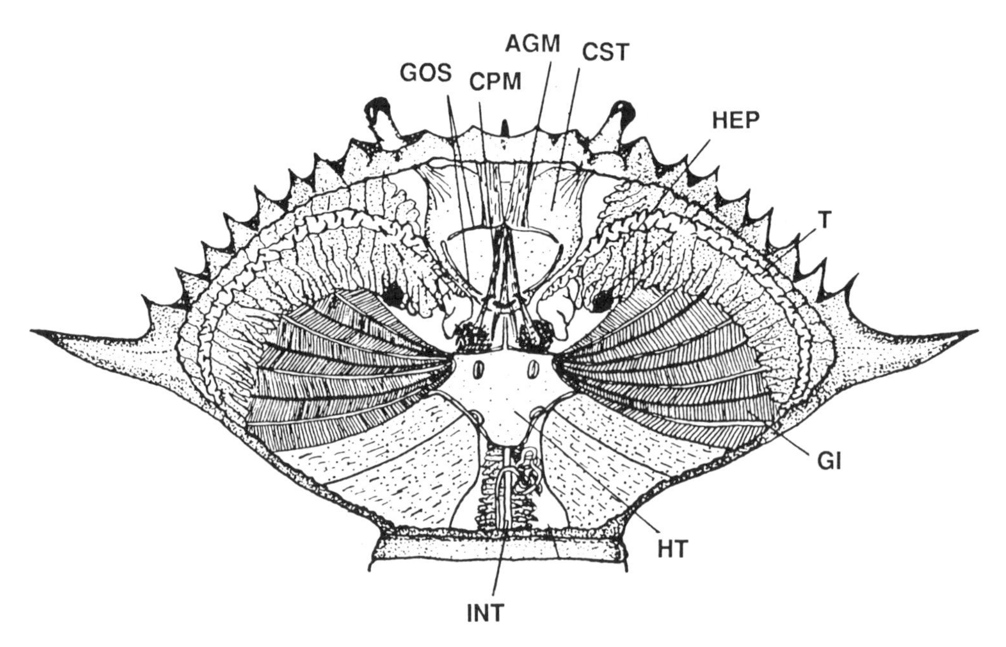

Figure 6. Dorsal dissection of the male as it appears after the carapace and hypodermis have been removed. AGM–Anterior gastric muscle; CPM–Cardiopyloric muscles; CST–Cardiac stomach; GI–Gill; GOS–Gastric mill; HEP–Hepatopancreas; HT–Heart; INT–Intestine; T–Testis.

by 2 and 1. Notice that a segment (the epipodite) of the third maxilliped extends into the branchial chamber. Notice the setae on this exopodite. They are responsible for cleaning the gills. There are five pairs of appendages associated with the head: 2nd and 1st maxillae, mandibles, antennae (second antennae) and antennules (first antennae). Note that the seminal receptacles are located ventrally on the 6th thoracic segment of the female.

Internal anatomy

1. Cut off the lateral spines on the carapace. This frees the outer membranes from the carapace so they don't tear and damage other internal structures.
2. Carefully cut a rectangle (3 cm x 4 cm) in the center of the shell beginning in back of the eyes and including the area above the heart (see Figure 6).
3. Carefully, using a sterilized scalpel or dissecting needle, slightly lift and free the attached membrane from the rectangle you've just cut. Be careful not to damage the heart under the carapace.
4. Now you can free the rest of the shell by going around the edge of the area exposed by the removal of the shell. Slip off the shell.

Locate, identify and list the function of each of the following:

- gills (dead man's fingers)
- cardiac stomach
- gastric mill (in stomach)
- gastric muscles
- intestine (behind heart and under connective tissue)
- heart
- testes (male)
- hepatopancreas (sometimes called mustard)
- ovary (size varies with seasons; female)

References

Crabs. In the water column. (1988, April). 2 (8). Gulf Coast Research Laboratory, P.O. Box 7000, Ocean Springs, MS 39564.

Hall, W.R., Jr. *Delaware's blue crab.* MAS Bulletin. University of Delaware Sea Grant Program, Newark, DE 19716.

Hicks, D.H. *Don't cook blue crabs by their color.* MAS Note. University of Delaware Sea Grant Marine Advisory Service, College of Marine Studies, Lewes, DE 19958-1298.

Pyle, R. & Cronin, E. (1950, August). *The general anatomy of the blue crab, Callinectes sapidus Rathbun* (publication no. 87). Solomons Island, MD: Chesapeake Biological Laboratory.

Rees, G.H. (1963, December). *Edible crabs of the United States* (fishery leaflet 550). Washington, DC: U.S. Department of the Interior, Fish and Wildlife Service, Bureau of Commercial Fisheries.

Sulkin, S.D., Epifanio, C.E. & Provenzano, A.J. (1982). *The blue crab in mid-Atlantic bight estuaries: A proposed recruitment model* (technical report) (publication no. UM-SG-TS-82-04). Maryland Sea Grant Program, College Park, MD 20742.

A Laboratory Study of Climbing Behavior in the Salt Marsh Snail (*Littorina irrorata*)

Steve K. Alexander

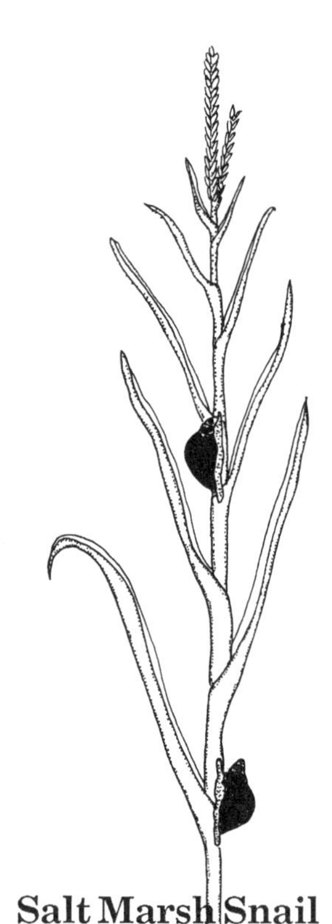

An animal in the wild receives stimuli from its surroundings (such as a mate within eyesight) which it converts into outward-directed activity (such as courtship). This outward-directed activity in response to stimuli is called behavior. While behavior in vertebrate animals is largely attained through experiences following birth (learned), that of invertebrate animals is largely inborn (instinctive). Instinctive behaviors provide some benefit to the animal, whether it is successful mating, procurement of food, or escape from predators.

A laboratory exercise dealing with instinctive behavior would help teach students about survival and adaptiveness of invertebrate animals in the wild. For such an exercise to succeed, the animal to be used must be readily available and must thrive in the lab. The animal also must display the desired instinctive behavior with a minimum of equipment and manipulation. Finding this combination by trial and error can be time consuming.

This chapter describes a laboratory exercise that fits all the above criteria for success. The exercise examines instinctive behavior (climbing) in the salt marsh snail, *Littorina irrorata*. I used this exercise during the 10 years I taught marine ecology at Texas A&M University in Galveston, Texas. The exercise will fit well in the laboratory portion of a secondary biology or marine biology course.

Salt Marsh Snail

The salt marsh snail, *L. irrorata*, is a common inhabitant of salt marshes along the East and Gulf coasts of the United States. In certain marshes, snail densities can reach 100 to 200 individuals/m^2. Densities of this magnitude correspond with lush growth of salt marsh grass, *Spartina alterniflora*. Density estimates are easy to make when water covers the marsh surface at high tide because snails are visible above the water line, clinging to *S. alterniflora* leaves. When the tide recedes and exposes the marsh surface, snails climb down grass stalks and feed on lower decaying portions of plant stalks, marsh sediment and algal mats (Alexander 1979). Upon being covered by water on the next incoming tide, snails climb up plant stalks to again resume their position on *S. alterniflora* leaves above the water line (Figure 1).

The first to study climbing behavior in *L. irrorata* was Bingham (1972). He determined that certain environmental stimuli, such as light and gravity, orient climbing when snails are covered by seawater. However, he did not address the purpose of climbing. Early speculation maintained that climbing out of water was due to a threat of drowning. Hamilton (1976) cast doubt on this explanation when he concluded that predator avoidance was a more likely explanation. For example, he observed numerous Blue crabs, *Callinectes sapidus*, on the marsh surface at high tide capturing and eating snails. Additional support for predator avoidance occurred when Bleil and Gunn (1978) demonstrated 100 percent survival of snails kept underwater in the laboratory for six weeks. Most recently, Warren (1985) demonstrated that snails in the field artificially held below the water at high tide suffered a substantially higher rate of mortality due to predation. Collectively, these studies suggest that climbing behavior in *L. irrorata* is an instinct to avoid predators. For additional information on classification, ecology and biology of *L. irrorata* and related snails (gastropods), refer to Barnes (1987).

Materials

To complete the exercises, each group of students must have the materials listed in Table 1. *L. irrorata* can be obtained from Gulf Specimen Company (P.O. Box 237, Panacea, FL 32346; 904-984-5297). Artificial seawater is available from local pet stores as

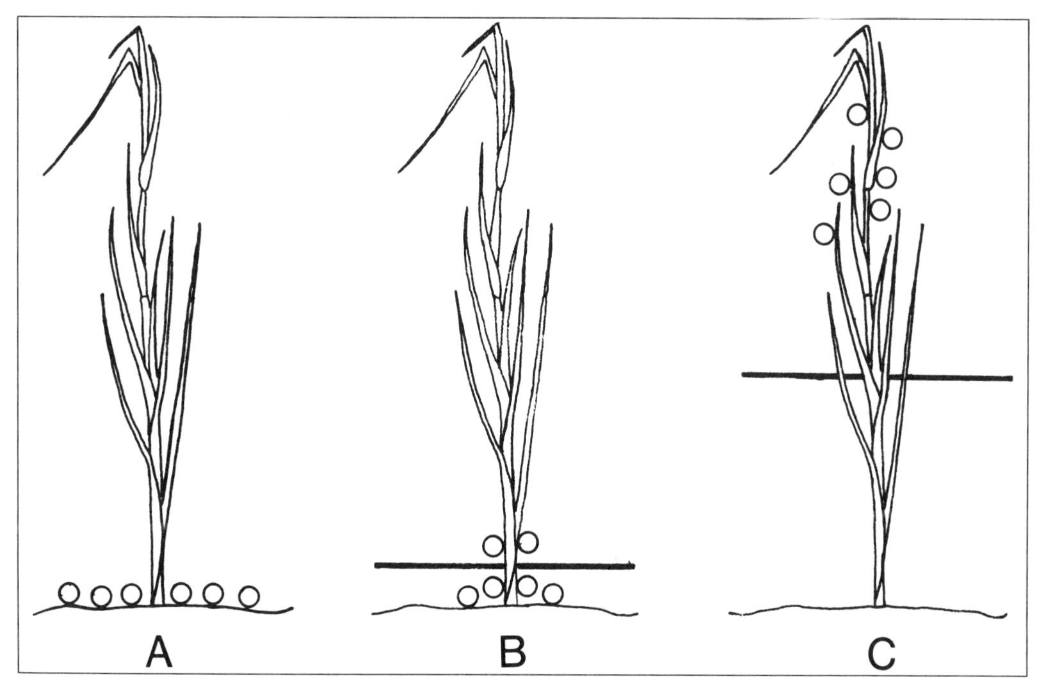

Figure 1. Position of snails (0) in relation to tide level. A. Low tide: snails feeding on marsh floor. B. Incoming tide: snails covered by seawater begin to climb stalks of marsh grass. C. High tide: snails on marsh grass leaves well above water line.

a dry salt to which water is added. Clear plastic tubes and white tube containers can be purchased at a local hardware store (sold as PVC pipe). If the clear plastic tubes are not available locally, they can be purchased from Forestry Suppliers (P.O. Box 8397, Jackson, MS 39204-0397; 800-647-5368). They are listed as replacement liners for hand-held corers.

Table 1. Supplies required (per student group) for all exercises described in the text.

Littorina irrorata * (20)
Beaker, 1 liter (2)
Seawater, artificial (2 liters)
Clear plastic tubes, 2" x 20" (2)
White tube containers, 20" (2)
Light source (1)
Miscellaneous items: cellophane, foil, rubber bands, marker, wire and centimeter ruler

*Keep snails in a foil-covered aquarium partially filled with artificial seawater until you're ready to use them.

Procedure

The three exercises below are designed to be done in sequence. In Exercise 1, students will observe climbing behavior in snails. In Exercises 2 and 3, students will examine the role of light and gravity, respectively, in climbing orientation. If sufficient time allows, I recommend students repeat each exercise two or three times to establish a pattern of response.

Exercise 1. Climbing behavior

Add 10 snails to the bottom of each of two beakers. Mark the side of each beaker at 5 cm intervals. Fill one beaker with artificial seawater up to the 10 cm mark. Cover both containers with cellophane and leave undisturbed on the countertop (movement of and around containers will alter the behavior of snails). With as little disturbance as possible, record the number of snails in each 5 cm interval every five minutes for 30 minutes. Students will observe the rapid ascent of snails up the side and out of water in the seawater-filled container (Figure 2).

Exercise 2. Orientation by light

Tightly seal one end of two clear plastic tubes with cellophane and rubber bands. Add 10 snails to the center of each tube and allow at least six to attach to the side of each before gently filling them with seawater (be careful not to dislodge snails).

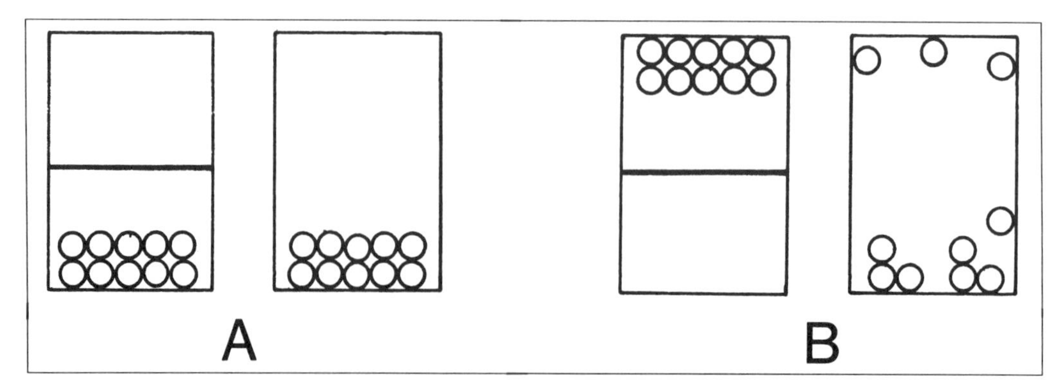

Figure 2. Response of snails to being covered with seawater. A. Position of snails (0) at zero-time. B. Position of snails (0) after 30 minutes.

Tightly seal the other end of the tubes as before and place them inside the white tube containers. Seal both ends of one tube container with foil (to eliminate light), but only one end of the remaining tube container. Place both tube containers side by side on a horizontal surface with the open end of the one tube container directed toward a light source. After 30 minutes, carefully remove the clear plastic tubes from the tube containers and record the position of all snails in each tube. Snails are positively phototactic after submergence, displaying movement toward light in the open tube and random, non-directed movement in the dark tube (Figure 3). In the salt marsh, sunlight overhead likewise directs snail movement upward on plant stalks.

Exercise 3. Orientation by gravity

Empty the clear plastic tubes and allow at least six snails to attach at the center of each tube. Gently fill the tubes with seawater and secure the open ends with cellophane and rubber bands as before. Place each clear plastic tube into a tube container and cover both ends of each tube container with foil to exclude light. Place one tube container in an upright (vertical) position and secure it with wire. Place the remaining tube

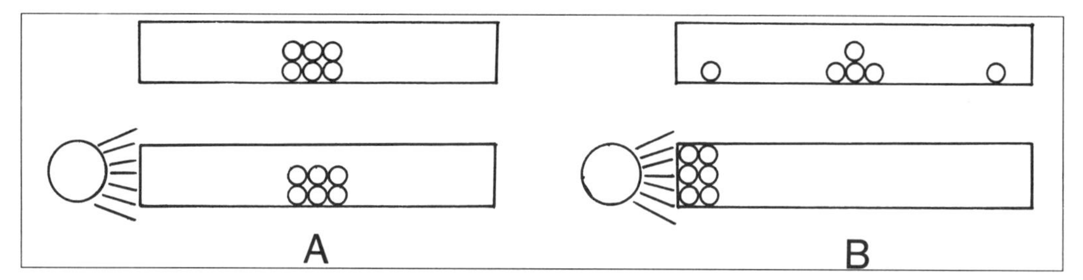

Figure 3. Response of snails to light after being covered with seawater. A. Position of snails (0) at zero-time. B. Position of snails (0) after 30 minutes.

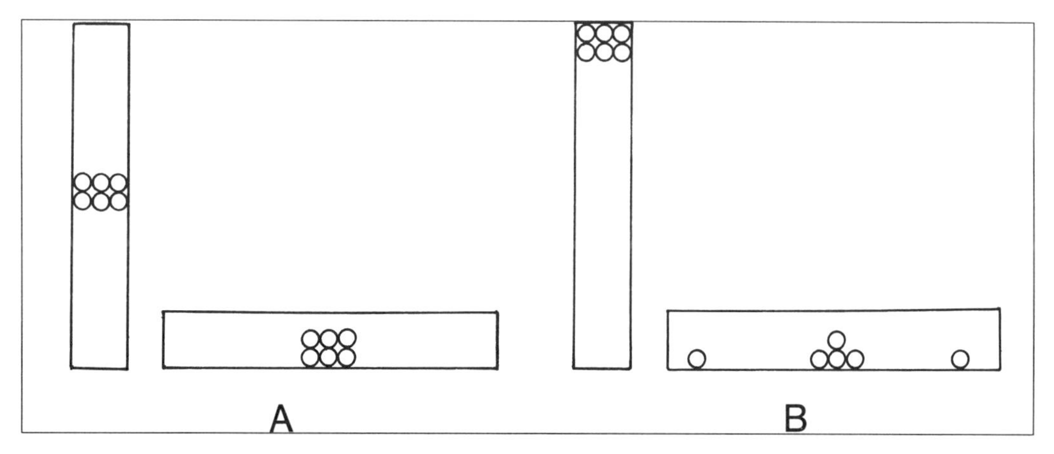

Figure 4. Response of snails to gravity after being covered with seawater. A. Position of snails (0) at zero-time. B. Position of snails (0) after 30 minutes.

container in a horizontal position nearby. After 30 minutes, remove the clear tubes and record the position of all snails in each tube. Snails are negatively geotactic when covered with seawater, displaying upward movement against gravity in the vertical tube and random, non-directed movement in the horizontal tube (Figure 4). In the salt marsh, movement against gravity directs climbing on plant stalks at night or on overcast days.

Report

The end product of this laboratory study is optional, but I recommend an oral or written report. Likewise, the content of the report is up to the individual instructor, but the following items deserve consideration:

1. presentation of observations (results)
2. interpretation of observations
3. comparison of observations with previous work on the subject
4. application to the animal's survival in the wild.

If a written report in the format of a scientific paper is desired, refer to Day (1988).

References

Alexander, S.K. (1979). Diet of the periwinkle *Littorina irrorata* in a Louisiana salt marsh. *Gulf Research Reports, 6*(3), 293-295.

Barnes, R.D. (1987). *Invertebrate zoology* (5th ed.). New York: Saunders College Publishing.

Bingham, F.O. (1972). The influence of environmental stimuli on the direction of movement of the supralittoral gastropod *Littorina irrorata*. *Bulletin of Marine Science, 22*(2), 309-335.

Bleil, D.F. & Gunn, M.E. (1978). Submergence avoidance behavior in the periwinkle *Littorina irrorata* is not due to threat of drowning. *Estuaries, 1*(4), 267.

Day, R.A. (1988). *How to write and publish a scientific paper* (3rd ed.). New York: Oryx Press.

Hamilton, P.V. (1976). Predation on *Littorina irrorata* (Mollusca: Gastropoda) by *Callinectes sapidus* (Crustacea: Portunidae). *Bulletin of Marine Science, 26*(3), 403-409.

Warren, J.H. (1985). Climbing as an avoidance behavior in the salt marsh periwinkle, *Littorina irrorata* (Say). *Journal of Experimental Marine Biology and Ecology, 89*(1), 11-28.

Marine Biology in Action–II

Marine biology is but one segment of a marine science course taught in the Greensboro, North Carolina, public school system, so time is limited for the study of invertebrates. Field trips to the coast allow Greensboro students to see marine invertebrates in their habitats. When the invertebrate section of marine biology is introduced, students do research, as well as a poster/model and an informational paper on the various phyla of invertebrates. They present their classmates with the information they have gathered using models or posters to show skeletal structures, other internal and external parts, the functional significance of each part and any other information of importance on their animal.

Since many of the major invertebrate phyla have freshwater representatives, the major exception being Echinodermata, landlocked students could study local lake, stream and river habitats for many of the invertebrate creatures, just as students near coastal areas have the opportunity to study marine invertebrates. Samples of echinoderms, though, could be ordered from a biological supply house.

Students are responsible for researching the characteristics of a particular phylum and then choosing a member from that phylum to present to the class by way of a poster or model. Each selected creature is to be identified on the poster/model by its correct scientific name, as well as common name, and an internal and external view drawn or made.

A one-page paper about the creature describes characteristics, lists scientific and common name, identifies habitat, indicates type of feeding and includes any unusual features. Students should consider the following questions as they research their animal

1. Why is this creature found in this particular habitat?
2. What adaptations help it survive in this niche?
3. Has the habitat been subject to changing environmental conditions that could have an impact on the creature?

When the designated research time is complete, students present their findings in phylogenetic order to classmates.

Some options for your class, depending on class make-up and size, are:

1. Have students set up model habitats using aquaria. Habitats might be dunes, salt marshes, tidal flats, rocky coasts, freshwater marshes and flats or rocky streams. Organisms should be a part of a typical habitat.
2. Print a student field guide booklet. Students select a typical creature in a chosen phylum and write up characteristics, including drawings of internal and external structures, habitats, type of feeding and the scientific and common name. Submitted materials are copied, compiled and one field guide is given to each

student. A student artist could design a cover.

3. Have students prepare worksheets to go along with their presentations. These can be crossword puzzles, search-and-find, acrostics, anagrams, fill-in-the-blank, etc., all using information related to their particular organism and its phylum.

4. Students submit at least five test questions, with answers, for use on a unit test (teacher revision suggested).

The example shows a typical marine science class sign-up that could be modified for appropriate freshwater phyla and animals:

Protista:	Radiolaria and Foraminifera and/or Dinoflagellates
Porifera:	Calcarea, Hexactinellida and Demospongiae skeleton types
Cnidaria:	Hydrozoa, Anthozoa (corals) and Scyphozoa (jellyfish)
Annelida:	Polychaeta (featherduster worms)
Echinodermata:	Asteroidea (sea star), Echinoidea (sea urchins and sand dollars) and Holothuroidea (sea cucumbers)
Arthropoda:	Xiphosura (horseshoe crabs), Crustacea (barnacles, copepods and decapods such as lobsters, shrimp and crabs)
Mollusca:	Polyplacophora (chitons), Gastropoda (whelks), Bivalvia (clams, oysters, scallops) and Cephalopoda (octopods and squid)

Selected miscellaneous taxa are considered as needed to have enough creatures to go around. These might include Ctenophora, Platyhelminthes, Nematoda, Bryozoa and Urochordata (sea squirts). Many of the phyla previously listed have typical freshwater representatives that could be adapted for areas not near the coast.

I assign grades for posters/models, for a properly written sheet of information, for worksheets and for the actual presentation of the material. Students also receive a grade if a unit test is given on the material .

Allowing for student preparation time, presentation of projects and testing, the unit will take two to three weeks, depending on the number of students and projects to be presented and how much various phyla might be subdivided to fit the needs of larger classes.

–Mary Ann Johnson

Songs of Giants: Bioacoustics in Cetaceans

Karen Travers

The sea is anything but silent. Clicks, snaps, squeaks, croaks, whistles, booms and moans resound throughout the ocean depths. From the grunts and growls of the ocean sunfish to the foghorn-like blasts of the toad fish, marine animals produce an interesting variety of sounds. Sounds facilitate mating, mark territory, aid navigation and promote group unity. The most extraordinarily complex and often hauntingly beautiful sounds are those produced by cetaceans: baleen and toothed whales. For the educator, the study of sound in whales provides an opportunity to capitalize on our natural fascination with this unique group of intelligent mammals while integrating both biology and physics in the classroom.

Whales use sound to communicate and "see" underwater. This provides a fine example of successful adaptation to environmental constraints. Sound communication in cetaceans evolved because it offered significant survival value. Below 150 meters the sea is dark, making vision almost useless; surface layers provide only minimal visibility. A sense of smell is of limited value since sea water is a poor conductor of scent particles. Sea water, however, is an excellent conductor of sound waves, making sound an efficient method of gathering information.

Sound Waves

Sounds are waves of alternating pressure changes that pass through a medium such as water or air and radiate from the source. To demonstrate this principle for students, drop a pebble in a pan of water and observe the concentric ripples. Sound waves radiate up and down as well. The water surface acts as a mirror, reflecting 99 percent of the sound back into the water.

Sound waves carry energy. The intensity or volume of this energy is measured in decibels (dB); human conversation, for example, is approximately 60 dB. The loudest animal sound underwater is the whistle of the Blue whale which can be detected 530 miles away when carried through underwater deep sea channels. Another important sound measurement is wavelength: the distance between one pressure peak and the next. Frequency is the number of waves that pass a point in one second. The complete wave cycle passing in one second is expressed as 1 Hertz (Hz). Low notes such as those produced by a bass violin or Blue whale are of low frequency, long wavelengths and are called infrasonic sounds. The high frequency, short wavelength pulses used by dolphins and bats to echolocate are called ultrasonic sounds. Humans can hear sounds between 20 Hz and 20,000 Hz (20 kHz). Dolphins are capable of producing pulsed clicks from .25 to 220 kHz, well beyond human hearing.

The greater density of water makes it a more efficient conductor of sound energy than air. Sounds actually travel at almost one mile per second in sea water, almost five times faster than in the atmosphere. If you have access to a pool, try an interesting sound conduction comparison. Have students mark off approximately 30 meters of the pool. While one person stands on the pool edge banging on a pot with a metal spoon, the rest of the group stands 30 meters away and listens. Repeat the process underwater. Do this several times and discuss the results. The class should be able to note that the sounds produced underwater reach our ears considerably faster than those traveling the same distance through air.

Humans actually hear poorly underwater because the air trapped in our ears acts as a plug or barrier. Cetaceans have solved this problem by conveying sound waves through the bony jaw to the inner ear. Unlike humans, whales lack vocal cords. Instead, air is forced through air passages in the head causing nasal plugs and other structures to vibrate and produce sound. When we consider that whales evolved 30 million to 50 million years ago, it is hardly surprising that their cerebral cortex has developed a sophisticated acoustical transmitting-receiving center ideally adapted to the marine environment.

Humpback Whales

Of all the animals, Humpback whales (*Megaptera novaeangliae*) have the most complex and varied vocalizations. Humpback songs consist of a series of sounds put together into a distinct pattern which is then repeated by the singer. Solitary males may sing their mysterious songs, containing between two and nine "themes," for hours. Each song is unique to the individual animal and specific to the region. Even more remarkable is that the "top tune of the year" changes progressively in time so that this year's song content may be very different from what is recorded a year later. Just why humpbacks sing is a subject for speculation. Since all the singers are male, they may be singing to attract mates, to demonstrate dominance in breeding territory, or perhaps to synchronize ovulation in breeding females.

A number of excellent whale recordings, including Roger Payne's classic "Songs of the Humpback Whale," are available through major biological supply catalogues and in many nature bookstores. Play one of these recordings and have the students try to determine the mood or intent of the animal making the vocalizations. If your recording has the calls of more than one species, try comparing the vocalizations for complexity, pitch and duration. Beluga whales (sometimes known as "sea canaries"), Right whales, Grays and Orcas all have distinctly different calls. As the class listens, have them draw the high and low sound variations as one continuous line on a piece of paper. The results will look very similar to sonograms or voice pictures scientists record using underwater microphones called hydrophones.

Dolphins

Many marine mammals including dolphins, Sperm whales and some seals examine their surroundings using directed pulses of sound. Bottle-nosed dolphins (*Tursiops truncatus*) have been studied since World War II because of their small size, intelligence, trainability and highly developed system of echolocation. In addition to their Morse code-like pattern of clicks and whistles, barks and whines which they use for communication, these small-toothed whales produce intermittent pulses of ultrasonic vibrations in their air passages. The sounds are "focused" by the fatty, bulbous melon located between snout and blowhole. They reflect off an object and return to the dolphin's lower jaw bone and the melon where they are transmitted by a series of bones to the inner ear. Simulate the clicks by running your fingernail over a comb and the whistles by allowing the air to escape from an inflated balloon while you stretch the mouthpiece taut with your fingers.

The use of echolocation by dolphins has been demonstrated by experiments in which their eyes have been covered while they navigate an obstacle course and find food. Dolphins are capable of discriminating among different kinds of fish and even between different objects of identical size and shape. This skill is all the more remarkable because sound waves bend or refract as they pass through the various ocean water layers. If either the dolphin or its prey is moving, the resulting shifts in echo frequency will give specific information about prey location and speed.

A dolphin increases the sound pulse rate to as high as 400 pulses per second as an object is approached and reduces as it is passed or eaten. A dramatic demonstration of this can be heard on the recording "Sounds of Sea Animals"(1959). Listening to this demonstration, it is easy to see how dolphins can use these loud bursts of sound to stun or even kill their prey.

In order to distinguish individual components, it is necessary to slow up the sound. Have students listen on the same record to the gradual slowing of a dolphin's echolocation clicks to 1/64 of normal speed until the individual echoes can be distinguished. Ask students to think of some practical applications of this unique ability. A tiny echolocation device has been built into the eyeglasses of the visually impaired to help them get around safely. The Navy has been using dolphins to find and retrieve objects near underwater stations and labs.

Exciting experiments are being done with dolphins using signs, symbols and computerized sound equipment to find out just how sophisticated their system of communication actually is and how close they come to using language in the human sense. Undoubtedly, the intricacy of their songs, their ability to probe the bodies of their companions with sound to determine emotional state and the subtlety of their group communications may rival and even surpass ours in some ways. It is tempting to find higher meaning in cetacean sounds but we must be careful not to take these messages out of context and anthropomorphise them. The wonder and mystery of cetacean sound is fascinating enough on its own.

Extension Activities

1. Have students make up their own adaptation for communication with appropriate anatomical structures. Tell how it aids survival.
2. Visit an aquarium or other marine facility with live marine mammals to observe their behavior and communication first-hand. Ask if hydrophones are available.
3. Explore other methods of communication (e.g., body language, sign language, etc.). See how many methods can be used to convey simple messages such as danger or the location of food. What happens to some of these methods when the message becomes more complex?
4. Investigate other cetacean adaptations to the marine environment such as the ability to deep dive or live in frigid polar waters.
5. Research the history of whale sound in conjunction with human marine activity from the "singing mermaids" of Homer's day to the disruption of naval sonar in World War II.
6. Bats' use of echolocation in the atmosphere is the counterpart of dolphin echolocation in the ocean. Compare the anatomy, physiology, adaptive complexes and the problems air and water respectively impose on echolocation in both groups of animals.
7. Listen to a recording of "Callings" by the Paul Winter Consort and hear how musicians have responded to marine mammal sounds using the jazz mode. Have students experiment with their own creative expressions.

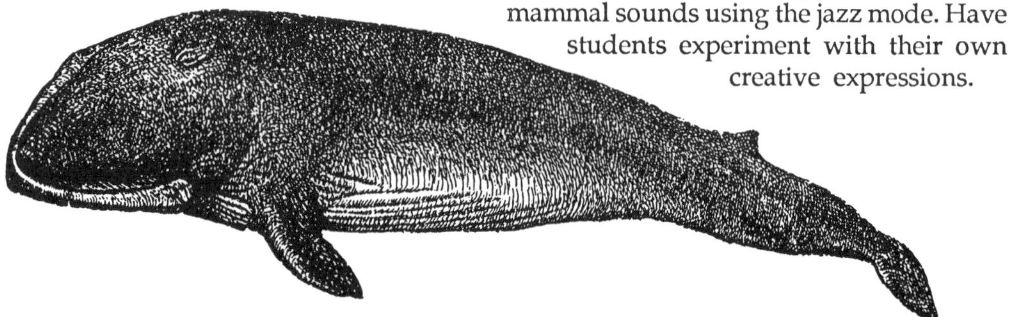

References

Bright, M. (1984). *Animal language.* Ithaca, NY: Cornell University Press.

Heintzelman, D.S. (1981). *A world guide to whales, dolphins and porpoises.* Tulsa, OK: Winchester Press.

Lerman, M. (1986). *Marine biology: environment, diversity, and ecology.* Menlo Park, CA: The Benjamin/Cummings Publishing Co.

Miller, F. (1980). *Concepts in physics.* New York, NY: Harcourt Brace Jovanovich.

Minasian, S.M., Balcomb, K.C. & Foster, L. (1984). *The world of whales.* Washington, DC: Smithsonian Books.

Slijper, E.J. (1962). *Whales.* Ithaca, NY: Cornell University Press.

Records

Deep voices: The second whale record. (1977). Hollywood, CA: Capitol Records.

Payne, R. (1970). *Songs of the humpback whale.* Hollywood, CA: Capitol Records.

Sounds of sea animals. (1959). New York: Folkways Science Series FX 6125.

Winter, P. (1980). *Callings.* Litchfield, CN: Living Music Records.

Fishy Business

Frances L. Lawrence

This activity simulates several state fisheries management programs over 10 seasons of fishing. During each season, teams calculate the effects on their fisheries of fishing effort, reproduction and recruitment rates, management decisions and unusual natural or man-made occurrences.

Objectives

After taking part in this activity students should be able to:

1. Use several basic vocabulary terms of fishery management.
2. Predict and calculate the effects given options have on stock size.
3. Analyze a fishery management situation, consider desired outcomes and select among options to achieve those outcomes.
4. Design and carry out a plan for making accurate calculations based on given percentages.

Materials

- 1 calculator (per team)
- 1 copy of Directions and Sample Data Sheet (per team)
- 1 set CHOICE cards and 1 set CHANCE cards

Procedure

Advance preparation

1. Have students read newspapers and magazines and listen to television and radio news for issues related to fisheries and their management.

Vocabulary

density dependent........factors dependent on the number of fish that regulate size of a population
density independent....factors not related to the number of fish that regulate the size of a population
harvest............................number of fish removed from the stock by fishing
fishing mortality...........fish removed from the stock by human actions
natural mortality...........fish removed from the stock by natural events
recruitment.....................number of young fish entering the fishery
stock ....a group of one species that spawns together, is taken together and can be managed as one unit

CHANCE cards begin with ☛

☛	☛	☛
Quantities of medical waste floating offshore; fishery closed for 3 weeks. Lose 10% of juveniles and 10% of adults.	Oil spill in your waters drifts south, fouls estuaries. Pass this card to state south of you. That state loses 20% of juveniles. Your state loses nothing.	Your fishery product is in great demand, prices high this season. Increase number of boats by one, two, or three this season.
☛ Coast Guard impounds some boats in violation of "zero-tolerance" policy. Decrease number of boats by two.	☛ Ideal conditions, recruitment rate higher than usual; increase number of juveniles by 10%.	☛ Climatic disaster: hurricane wipes out much of current year class. Reduce number of juveniles by 40%.
☛ Poachers take 20% of your mature adults at the beginning of the season, before you start fishing.	☛ Most juveniles in the fishery are spawned in your waters. To protect their stock long-term, other states push through a federal law prohibiting your state from taking any juveniles this season.	☛ Nothing out of the ordinary happens this year.
☛ Abandoned dams blocking rivers removed, historic spawning grounds reopened, recruitment increased by 10%.	☛ EPA eel grass planting successful; nursery ground increases recruitment by 10%.	☛ U.S. Fish & Wildlife Service restocking successful; recruitment increases by 10%.

CHOICE cards begin with ✔

✔	✔	✔
Poor catch this season. (A) Remove one boat or (B) Fish in another state. If B, pass card to another state, which must reduce stock by 5%.	Regulatory agency ignores your fishery. (A) Add or lay off up to three boats or (B) No change.	Some fishermen want to fish for giant squid. (A) Let them take two boats and go or (B) Fish as usual.
✔ Gear problems limit fishing to one size fish. (A) Catch only juveniles, (B) Catch only young adults or (C) Catch only mature adults.	✔ You have become rich and powerful. (A) Throw away the chance card you drew or (B) Rent up to five extra boats.	✔ You lose three boats to fire of suspicious origin. (A) Accept insurance and replace boats now or (B) Investigate, postponing replacement till next season.
✔ Your controversial new gear has just been made illegal. (A) Fish anyway, 10% increase in harvest or (B) Remove, it, 10% decrease in harvest.	✔ Sea Grant offers efficient new gear to your fleet for field testing. (A) Accept it, doubling per boat catch or (B) Say "no, thank you."	✔ Vessel safety is a problem; one of your boats is questionable. (A) Lay it up for overhaul or (B) Fish as usual.
✔ You have a chance to work two extra boats in area closed due to pollution. (A) Fish the two boats or (B) Refuse this unethical opportunity.	✔ You catch a boat from the state north of you fishing in your waters. (A) Impound boat; you get boat and its fish, other state loses both or (B) Do nothing.	✔ Foreign fishing encroaching. (A) Let them take 5% of juveniles or (B) Send them to another state where they take 5% of juveniles.

2. Introduce vocabulary terms and discuss definitions. Using the vocabulary terms, discuss the news reports, eliciting possible management options.
3. If necessary, have class review the mechanics of making calculations and solving word problems with percentages.

Instructions for activity

1. Divide the group into teams of three to five each. Consult a map to find an area that has as many contiguous states on the same body of water as you have teams. Assign one team to represent each one of the states.
2. Review directions with the class. The objective of the game is to stabilize the fishery so the harvest remains relatively constant each year (thereby encouraging stable prices and steady employment of fishermen) while the stock is maintained without overpopulation or stock depletion.
3. After 10 seasons, teams compare their fisheries, management strategies and results in class discussion.
4. The teacher appoints a five-member Fishery Management Council which selects the best managed fishery, based on its review of each state's records.
5. Evaluation: Have students individually design original fishery management plans based on a new, but similar situation of your choice.
6. Extension activity: Students research actual fishery management regulations for the states they represented. Discuss in class, identifying similarities and differences among states.

Directions

1. Assume the following:
   - All adults reproduce, juveniles do not and the natural life span of each fish is three seasons.
   - All adults and juveniles are available for harvest.
   - Each state begins with 500 mature adults, 1,000 young adults and 4,000 juveniles.
   - Each state begins with 10 fishing boats, each of which catches 100 fish per season, and adds one boat to its fishery each season. The number of boats may be further increased or decreased by CHOICE or CHANCE.
   - Between the end of one season and the beginning of the next:

     a) 50 percent of juveniles and young adults die (50 percent mortality)
     b) surviving juveniles become young adults which then spawn, producing two juveniles each
     c) Surviving young adults become mature adults which then spawn, producing four juveniles each
     d) 100 percent of mature adults die

- If, at the end of any season, the total population exceeds 20,000, natural, density dependent mortality sharply increases to 75 percent among juveniles.

2. During each of the 10 seasons, the following occur:

- Each team draws one CHOICE and one CHANCE card. They must do as the CHANCE card directs and must choose what they believe to be the best management option presented on the CHOICE card. Return all cards to the decks and shuffle after each season. CHOICE/ CHANCE instructions are for the current season only.
- Make calculations for the season. The method and format for making computations and recording data is up to each team, but must:
  - a) proceed in the order shown on the sample data sheet, left to right;
  - b) not violate any assumptions or CHOICE/ CHANCE requirements; and
  - c) be mathematically correct.

  All teams move to the next season at the same time.

3. At the end of 10 seasons, each team must have harvested a total of at least 100,000 fish.
4. Compare your state team's results with those of other teams. Inspect their records. Which seems to have the most stable fish population? If you were a fisherman, in which state would you prefer to work? Why? Which management policies do you think work best? Which do you think is more important: CHOICE or CHANCE? How does this simulation differ from actual fishery management situations?

SAMPLE DATA SHEET

STEPS:

① Draw CHOICE and CHANCE for instructions for first season
② Calculate recruitment for the season
③ Record fishing mortality
④ Record numbers of fish in stock
⑤ Calculate natural mortality
⑥ Draw CHOICE or CHANCE for instructions for next season
⑦ Calculate numbers of adults which will spawn during new season.
At end of season 1, young adults become mature adults for season 2, juveniles become young adults, and mature adults die.

Repeat steps ② - ⑦ for remaining seasons.

EXAMPLE:

	MA	YA	J(MA) + J(YA)	TOT
① Season 1				
② RC	*500*	*1000*	*2000 + 2000*	*5500*
③ FM	*-500*	*-250*	*-250*	*1000*
④ ST	*0*	*750*	*3,750*	*4500*
⑤ NM		*-375*	*-1875*	*2250*
⑥ Season 2	↙		↙	
⑦/②	*375*	*1875*	*3750 + 1500*	*7500*
③	*-300*	*-200*	*-600*	*1100*
④	*75*	*1675*	*4650*	*6400*
⑤	*-75*	*-831*	*-2325*	*3237*
⑥ Season 3				

(Continue as before)

The figures show one possible way to manage the fishery through two seasons. This is not necessarily the best way, nor does it reflect any adjustments for CHOICE and CHANCE.

KEY

MA = mature adult
YA = young adult
J = juvenile
RC = recruitment
FM = fishing mortality
ST = stock size after harvest
NM = natural mortality
TOT = totals

Pieces & Parts of the Environment, an Introduction to Resource Partitioning

Barry W. Fox

Like the pieces of a puzzle, natural resources in the environment are divided and shared by many organisms. Put all the pieces together and we see how nature is actually a web of animals and plants depending on each other for survival. Their adaptations allow them to use and share available resources in many ways. This sharing or dividing of resources is called "resource partitioning." The term partition means "to divide" and the environment is actually divided among its inhabitants.

Both marine and freshwater environments offer excellent examples of resource partitioning. Although the organisms and exact mechanisms may differ, the process and final results are the same in both environments. This chapter will examine the use of food resources by some freshwater fishes.

Feeding specializations are an important mechanism of resource partitioning in a fish community. Different populations (species) can be divided into distinct feeding guilds (groups) using other fish, shrimp, aquatic insects, crustaceans, gastropods and plankton as major prey items. Resource partitioning may also occur along several axes including time and location of feeding, prey size, feeding methods and ontogeny (development) of feeding patterns. The result is a division of food resources to accommodate a variety of different fish species.

A typical warm, freshwater habitat may include pickerel, catfish, several species of sunfish, minnows, suckers and other species. The differences in preferred foods and feeding strategies allow different species to live in the same habitat and share food resources. Analysis of feeding habits reveals distinct feeding guilds that might appear as listed in Table 1.

Although pickerel and catfish both feed on other fish, catfish are nocturnal bottom feeders while pickerel are diurnal surface

Table 1. Feeding guide.

Fish Species	Dominant Prey
Adult pickerel, catfish	fish
Adult warmouth sunfish, large mouth bass	fish, crayfish
Adult blue spotted sunfish, pumpkinseed sunfish	snails, midge larvae
Adult mosquito fish, minnow, young chub suckers	small crustacea, insect larvae

or mid-water feeders. Because of their size differences, adult warmouth and blue spotted sunfish take smaller prey than their respective counterparts, largemouth bass and pumpkinseed sunfish. Spatial segregation can also be seen in the surface feeding mosquito fish, mid-water feeding minnow and bottom feeding young chub suckers. Sharing resources through time and space reduces competition between species and allows a more even distribution of available resources.

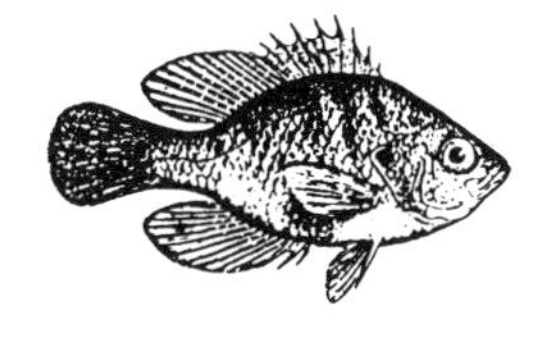

The ontogenetic development of feeding patterns as fish mature allows the use of a greater variety of foods. For example, very young pickerel feed on aquatic insects, juveniles feed on small fish and shrimp, and adults feed primarily on other fish. Young pumpkinseed feed mostly on midge larvae, but more on snails as adults. Similarly, young chub suckers prey on small crustaceans while adults feed on finger clams. Although many fish tend to specialize on certain prey, some fish, such as the pirate perch, eat a variety of prey items during all life stages.

The most common indicator of resource differences is the size and type of feeding structures related to prey. Certain species of fish may appear similar, but are separated by structural feeding adaptations. For example, the size and structure of the mouth influences efficiency in foraging on different prey. The small protruding mouth of the bluegill allows predation on small aquatic insects, while the larger mouth of the redbreast sunfish affords selection of larger insect prey. The large, extensible mouth of the largemouth bass and warmouth sunfish enable both species to use suction in capturing large prey, while the pickerel uses well developed teeth to capture and hold prey. Fish may also have a variety of tooth plates in the mouth for handling and rending prey. In addition, the location of the mouth facilitates surface, mid-water, or bottom feeding. These and other structural adaptations enable species to partition resources more efficiently.

The function of resource partitioning more clearly defines the role of each species in its habitat and facilitates in providing a niche. Each species has a particular set of resources it requires (food, space and shelter) that makes up its niche. No two species can occupy the same niche, however there can be varying degrees of overlap. The efficient division of resources allows greater niche and species diversity for a given habitat.

Game: PARTITIONS

The following activity demonstrates the function of resource partitioning in the aquatic environment. Participants will select feeding strategies and prey and then will collect available food resources in the environment. The activity can be modified according to the age and size of the group. There is no winner or loser, and students should not physically compete for prey cards. In this activity, all participants receive their share of resources.

Objectives

Students will learn how resource partitioning functions in the natural environment.

Materials (for 30 students)

- 30 paper bags
- 30 pencils/pens
- 30 resource cards (photocopy or create your own)
- 100 of each prey card: large and small insect, large and small fish, large and small crustacean, plankton (photocopy or create your own)

(Small, colored wood chips can be used for prey cards, different colors designating different prey. For young children, duplicate pictures of prey animals for ease of identification.)

Introduction

Introduce students to the concepts of habitat and resources (food, water, shelter). Ask students to give examples of various aquatic habitats and some organisms that may be found in each. Ask students to list the natural resources these organisms must have to survive. Introduce the concept of resource partitioning and give examples as described in the body of this chapter. Ask students to give examples of resource partitioning for resources other than food (e.g., space, nesting sites, territories, etc.).

Introduce the concept of the niche. Select one common aquatic organism and ask students to describe its niche (food, shelter, space, activity periods, nesting sites, etc.). Ask students to describe what happens when two species try to occupy the same niche (only one species can occupy a particular resource).

Resource Card

FOOD ITEM	LARGE SMALL	SURFACE FEEDING	BOTTOM FEEDING	PLANT FEEDING
NIGHT FEEDING				
DAY FEEDING				

Procedure (large open room)

1. Hand out one paper bag to each student.
2. Hand out resource cards or ask students to create their own (see attachment).
3. Identify three corners of the room as surface feeding, bottom feeding and plant feeding areas. Make labels on sheets of paper and place them at the appropriate corners.
4. Place one-third of the insect, fish, crustacean and plankton prey cards in each corner. Mix the large and small cards equally. Each corner should have about 33 of each type of prey card. For smaller or larger classes, use more or less prey cards. (Plankton prey are of one size only.)
5. Each student will play the part of a predatory fish in an aquatic environment. First, ask students to select to feed at night (nocturnal) or day (diurnal), then circle the appropriate word on the resource card. Make sure students do not all mark the same thing. This applies to the following selections as well.
6. Ask students to select their feeding location (surface, bottom, plants) and circle it on their resource card.
7. Ask students to select their prey (insect, fish, crustacean, plankton) and prey size (large or small). Have them fill this in on the appropriate blanks on their resource card. If plankton is selected, no size need be given.

Prey Cards

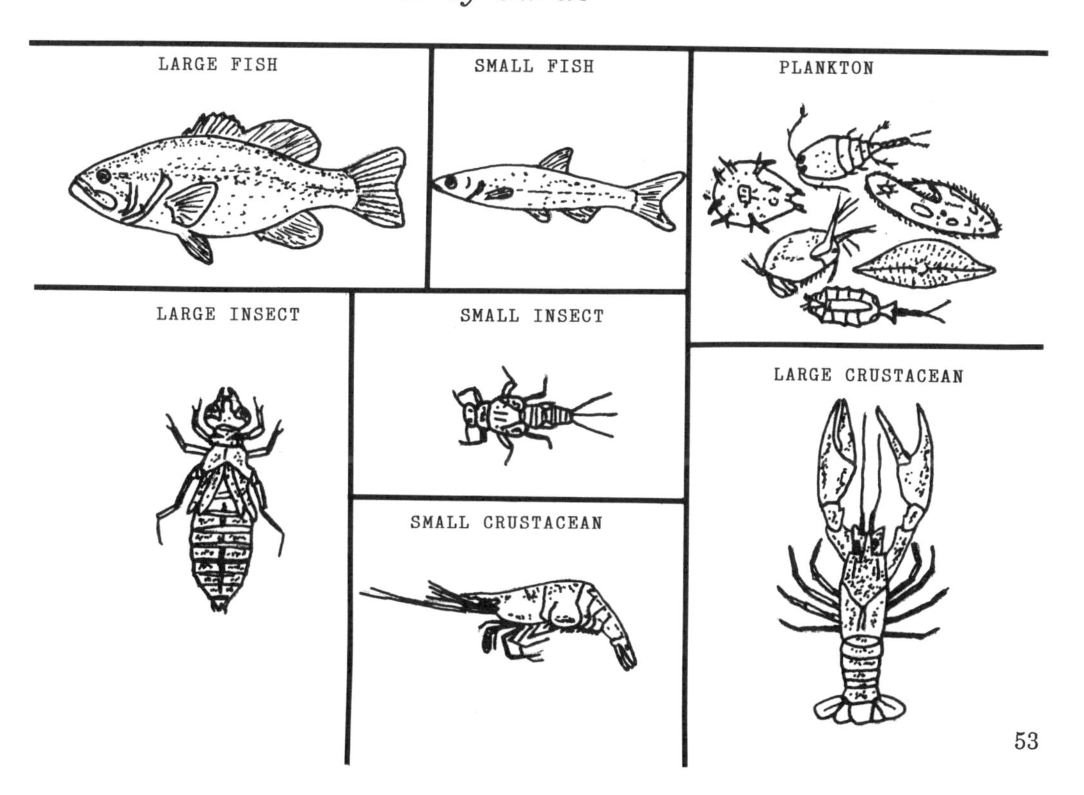

8. Each student is now prepared to feed on a particular size prey and at a particular time and place. The paper bag represents the predator's territory and stays in one place. Students go to the appropriate feeding areas, pick up one prey card of the correct type and take it back to their territory. They make a slash mark in the appropriate box on their resource card indicating the prey has been eaten, place the prey card in their bag, then return to "catch" another prey.
9. This is not a race. Students can take their time, act out the part of a predator and make all kinds of munching sounds as they "eat" their prey.
10. Begin the activity by stating it is 6 a.m., the nocturnal feeders are in their territories resting, the sun is rising and the diurnal feeders are starting to feed. These individuals then go to their appropriate feeding stations and "catch" one prey at a time. Quickly count off the hours of the day (for about a 30-second period) then announce that the sun is setting, it is 6 p.m., diurnal feeders are returning to their territories and it is now time for the nocturnal feeders to catch prey. Alternately give each group 30 seconds to feed.
11. Repeat the feeding rounds until supplies of prey cards are low. At this point, end the activity and discuss the results (step 14) or continue as follows.
12. When food resources become limited, animals often change their diet or feeding location. Ask students to repeat steps 6 and 7, selecting a new prey of the same prey size or a new feeding location. (Students will probably note which area has the greatest food supply.) Students cannot change their selection of diurnal or nocturnal feeding.
13. Repeat the feeding rounds as described in step 10 until food resources are consumed.
14. Ask students some of the following questions for review:
    a) From the prey items selected, what species of predatory fish might you have represented (bass, sunfish, pickerel, catfish, minnow, etc.)?
    b) How were the food resources divided in this activity (time, location, type, size)?
    c) What happens if there are too many of the same predator species (competition for food resources)?
    d) How do animals reduce competition when food resources become limiting (change diets, feeding locations)?
    e) Do you think there are food resources in the natural world that may be underused? (Some organisms are not heavily preyed upon but are an important part of the ecosystem.)
    f) This was a rather simple representation of resource partitioning. Discuss the complexities of involving 50 different organisms in this food web.

g) Have students count the number of prey items "eaten" and discuss how much is sufficient. The leader can make some arbitrary judgments on the numbers required for survival. (Example: to survive, a predator would need four of each large prey, eight of small prey, ten plankton, etc.)

Follow-up Activities

1. Have students thoroughly research an animal of their choice to learn about its natural history.
2. Repeat the game but add a pollution factor (a marked prey card which represents some type of contamination), then discuss possible effects on the food web.
3. Repeat the game but assign feeding strategies to students to make sure there are too many of one type of predator all seeking the same prey. Discuss the results.
4. Have students research and discuss the following ecological concepts: food chain, food web, feed pyramid, holding capacity, adaptive radiation. Relate these concepts to resource partitioning.

References

Ball, R.C. (1948). *Relationship between available fish food, feeding habits of fish, total production in a Michigan lake* (Technical Bulletin 206, pp. 1-59). Detroit: Michigan State College, Agricultural Experimental Station.

Dodson, S.I. (1970). *Complimentary feeding niches sustained by size selective predation.* Condensed from doctoral thesis, University of Washington.

Fox, B.W. (1982). *Habitat utilization and resource partitioning by juvenile fishes in a freshwater creek-marsh habitat.* Unpublished master's thesis, Virginia Commonwealth University, Richmond.

Gatz, A.J. (1979). Community organization in fishes indicated by morphological features. *Ecology, 60,* 711-718.

Ivlev, V S. (1961). *Experimental ecology of the feeding of fishes.* New Haven, CT: Yale University Press.

Keast, A. (1965). Resource subdivision amongst cohabiting fishes in a bay, Lake Opinico, Ontario (Pub. no. 13:106-132). Proceedings of the 8th Congress on Great Lakes Resources, Great Lakes Resource Division, Institute of Science and Technology.

Kushlan, J.A. (1976). Environmental stability and fish community diversity. *Ecology, 57,* 821-825.

Schoener, T.W. (1974). Resource partitioning in ecological parameters. *Science, 185,* 29-39.

Decision Making Through an Estuarine Systems Issues Simulation

Emmett L. Wright
David A. Wright

An estuary is a semi-enclosed body of water where rivers meet the ocean. The dynamic interactions between fresh and salt water provide unique habitats for a variety of organisms. The estuaries (bays) of the United States' coastline are facing deteriorating water quality and productivity. There are many different opinions of how estuarine resources should be used, and the complex issues that arise have been debated by citizens, agencies, industry and municipalities.

It is the biology teacher's responsibility to assist students in evaluating the known facts, not from just a scientific viewpoint but from a multidisciplinary perspective that takes into account historical, political, social and economic factors.

Our experience is that presenting estuarine issues information in a lecture format limits student exploration of the topic. Students generally are not motivated to ask probing questions or offer serious debate. Their main concern is to repeat "respectable" responses on the examination. We have tried and found successful a simulation strategy that permits both secondary and college students to debate the most and least critical issues.

Simulation

Bay Issues Simulation Exercise (Figure 1) asks students to rank a series of issues associated with the fictitious "Beautiful Bay" which are common to most estuarine systems in the U.S. (**Note:** The information to the left in Figure 1 is not initially given to students). One way to begin the simulation is to present a short lecture defining each Beautiful Bay issue; students then rank the issues as a homework assignment. Collect the assignment during the next class period and divide students into groups of three to five. Have each group develop a consensus of their individual rankings (see references for background.)

Develop Consensus

To encourage critical thinking, require the students to write out reason(s) for their selection of the two most critical and two least critical issues. Stress that credit will be given only for a documented response, not for hearsay or personal opinion. Encourage students to use the school library; some good references are listed at the end of this chapter. During the next class period, collect the second assignment and ask each group to select a representative to debate the importance of each Bay issue with representatives from the other groups.

Debate Choices

It should be noted that enthusiasm for and commitment to specific issues will create competition and spirited debate. Because of their involvement in the decision making, expect input from individual students in the audience. Be sure to establish a protocol for recognizing comments from the floor.

The issues can be debated at length by students. Require them to apply facts and principles and use scientific reasoning in ranking the issues. The motivation will remain high and students will be enthusiastic as they learn about the problems associated with good management of estuarine systems.

Students need to understand the difference between decision making based on the "facts" and decision making influenced either partially or wholly by value-laden opinions.

Be prepared to deal with students' value systems. Students need to understand the difference between decision making based on the "facts" and decision making influenced either partially or wholly by value-laden opinions. After the exercise have students assess their personal values and the values of their parents in terms of what they viewed as important in the debates.

As an alternate debate process, assign each group to a special interest group (I.G.). Possible groups include:

- Wetland users I.G. (commercial trappers, crabbers, clammers and oystermen)
- Recreation I.G. (boat owners, fishers and hunters, tourists)
- Developers I.G. (bankers, Realtors, homebuilders)
- Scientific I.G.
- Small Business I.G. marine owners, farmers, gas station operators)
- Industrial I.G. (steel company, shipping line, power company)
- Environmental Protection/Conservation I.G.

Each group is asked to rank and debate the issues from the perspective of its special interest. Follow this up by having a representative from each group debate the issues before the entire class. Role playing adds additional perspective to the complexity of the issues by reflecting the attempts of various elements of society to develop public policy to use, "save," or manage the bay. Again, students need to be aware of the influence of values and opinions. In role playing students are to assume the perspective of members of a particular interest group; this may be hard for some to do. Afterwards, examine the interest groups' values in relationship to the personal opinions.

Establish Priorities

It is not an objective of the activity for students to come up with the "correct" answer, particularly since the scientific and political communities cannot fully agree on priorities. Numerical rankings need not be required after the initial activities. Students can simply designate issues as either having a high or low priority. It is not essential that students agree. It is important that their priorities reflect careful thought and can be supported by scientific literature.

Develop a Balanced Picture

The major simulation goal is to develop a thorough analysis of each estuarine issue, to examine the trade-offs one could expect if a particular issue is or is not given top priority. Following class debate–which hopefully has led to some documentation and consensus–we typically hand out the information illustrated to the left of the issues in Figure 1. This input may encourage students to reconvene the debate, modify their choices, or question the teacher's sanity!

The simulation process illustrates the complexity of important issues associated with decision making for a major ecosystem. There is no clear cut decision-making or problem-solving process. Students will call on not only biological and other scientific information but on ideas from agriculture, economics, history, law, political science and sociology. This process could be adapted for students of other biologically-related issues such as farm management choices (soil and water conservation), birth control options, freshwater and land use issues, alternative energy sources (renewable/nonrenewable) and perhaps for important decision making about the introduction of a new biotechnology.

References

Chesapeake: The twilight estuary (film). (1985). College Park, MD: Maryland Sea Grant College. Maryland Sea Grant College, H.J. Patterson Hall, University of Maryland, College Park, MD 20742

Estuary (film). (1976). National Oceanic and Atmospheric Administration, Film Order Department, 12231 Wilkins Ave., Rockville, MD 20852.

Horwitz, E.L. (1984). *Our nation's wetlands.* Washington, DC: U.S. Government Printing Office.

Nybakkan, J. (1982). *Marine biology: An ecological approach.* New York: Harper & Row Books.

Teal, J. (1969) *Life and death of a salt marsh.* New York: Ballantine Books.

U.S. Environmental Protection Agency. (1982). *Chesapeake Bay: An introduction to an ecosystem.* Washington, DC: U.S. Government Printing Office.

U.S. Environmental Protection Agency. (1984). *Chesapeake Bay: A framework for action.* Annapolis, MD: EPA Chesapeake Bay Program.

Warner, W.W. (1976). *Beautiful swimmers.* New York: Penguin Books.

Wright, E.L. (Ed.). (1985). *Decision making/the Chesapeake Bay.* College Park: MD: University of Maryland.

Figure 1. Bay Issues Simulation Exercise

The "Beautiful Bay" is a vast natural and social resource that is deteriorating in water quality and productivity. Along with its tributaries, the Bay provides a transportation network on which much of the economic develoment of the Beautiful Bay region has been based. The Bay area provides a wide variety of water-oriented recreational opportunities such as boating and fishing and is a home for numerous fish and wildlife, a source of water supply communities and industries and is the site for disposal of waste products. Unfortunately, problems arise when people's intended uses of one resource conflicts with the natural environment, the use of another resource, or a different use of the same resource. Planning is needed to provide efficient and environmentally sound use of the Bay's resources.

You are one of several individuals up for appointment as the director of the Tidewater administration of the State Department of Natural Resources. One of the director's responsibilities is to develop a management plan for the Bay. A concerned organization of citizens groups, COCG, wishes to evaluate candidates' priorities concerning the important public issues facing the long-time management of the Bay's resources.

The COCG evaluation "quiz" is as follows: You must numerically rank the following Bay issues that must be addressed from that which is most critical in terms of having a healthy bay in the future to that which is the least critical. Your decision should take into acccount not only environmental concerns but also the social, recreational, economic and educational needs of Bay citizens.

---

*Recreational boating is second only to the major port of New City in the state's marine-minded economy. Since most marinas are filled to capacity, more public launch facilities are needed. Some environmental problems include oil-spill pollution, channel dredging, shore erosion, turbidity, safety and congestion.

_____ Unlimited recreational boating and marina development.

Wetlands play a key role in the state's estuarine environment. They provide basic nutrients to the food chain and habitat for many fish and wildlife species, as well as protect water quality, give flood protection and help control shore erosion. Tidal wetlands are protected by state law to a degree, but there is still considerable pressure to alter (fill in or develop) these wetlands. Freshwater (non-tidal) wetlands provide valuable habitat and food, particularly to waterfowl and fur bearers. These communities serve as buffers for storm erosion, aquifer recharge areas and filters for sediments and pollutants. Agricultural drainage, urban development and many other activities threaten non-tidal wetlands.

_____ Use of tidal and non-tidal wetlands.

Erosion of more than four feet per year threatens about 140 miles of the Bay's shoreline. Commercial shipping activities on the Bay and building on the shore can increase erosion, damage oyster beds and cause much greater erosion damage to developed shorelines than normal weather conditions. Most Bay beaches are on private land. As new public beaches are created, there is a demand for transportation routes to provide access; this can damage natural resources such as wetlands.

_____ Shore erosion and increased beach access.

---

* This rationale section is given to students only after consensus is reached.

The Bay shorelines are increasingly popular for large- and small-scale residential development. Negative impacts occur when facilities' demands (such as sewage treatment plants, police, schools, fire protection) exceed the areas' capacity to pay for them. Sedimentation, nonpoint pollution and loss of valuable habitat occur with growth of any size.

_____ More residential development.

The impact of sewage facilities on shellfish can be severe when sewage harms the water quality (oxygen, nutrients and residual chlorine, for example). New sewage treatment plants also increase high-density development because more treatment capacity is available.

_____ Increased number and capacity of sewage-treatment facilities.

The Port of New City generates about 15 percent of the gross state product and more than 300,000 jobs. Environmental problems include oil spill pollution, shore erosion and safety. Other toxic materials, when transported, also threaten the Bay. Dredging is a problem but must be done to keep ship channels open. The spoil (materials removed from the channel) must be disposed of safely so as not to endanger shellfish beds or other resources. Containment sites for the spoil are difficult to find and maintain.

_____ Enhanced commercial shipping and ports.

Fish and shellfish from the Bay are a major part of the state's recreational and commercial life. Overfishing, agricultural runoff, sewage stormwater discharge and industrial discharge all threaten these living resources. Moratoriums on the taking of striped bass and shad are already in effect, and the harvesting of oysters and clams is restricted in some areas because of contamination.

_____ Continual harvesting of living aquatic resources (fish and shellfish).

Much of the Bay's drainage area is agricultural–more than 100,000 farms are located there. Increasing home-development pressures encourage building on prime farm lands. Agricultural runoff of sediments, pesticides and fertilizers is an important source of nonpoint pollution in portions of the Bay and its tributaries. Timber is an important crop. Forests also serve as natural buffers, watershed protection, wildlife habitat and recreation areas. Most Bay-area forests are privately owned. Environmental problems such as erosion and sedimentation occur when poor practices are used in managing and harvesting timber.

_____ More intense use of decreasing acreage of agricultural and forested lands.

Industrial parks average more than 300 acres and provide facilities for several types of industries. The parks have great economic importance to the area because they provide jobs and tax revenue. Industrial parks also have great environmental impact on certain kinds of activities and the surrounding area.

_____ Additional industrial parks.

Demonstrating the Nitrogen Cycle in a Marine Environment

Kevin T. Patton

One of the essential aspects of marine biology is the concept of seawater as a matrix for communities of living organisms. Since we are not marine organisms ourselves, we need to develop an appreciation for the unique characteristics of the principle component of the ocean environment: seawater. It is comparatively easy to analyze for the relative concentrations of the various components of marine waters, but it is difficult to determine the nature of the complex chemical cycles that occur within these waters. One of the most critical pathways of chemical cycling in seawater (yet one that can be clearly understood by the beginning student) is the nitrogen cycle. A simple technique is available to the typical school laboratory for a comprehensive demonstration of the nitrogen cycle in the marine environment.

The marine nitrogen cycle can be summarized into the following steps:

1. Food materials, especially proteins, are metabolized by animals such as fish and by microbial decomposers resulting in the excretion of nitrogenous waste products, principally ammonia. Ammonia (NH_3 and NH_4^+) is toxic to most organisms and so must not acculumate in any great amount.
2. Nitrifying bacteria (genus *Nitrosomonas*) and other organisms throughout the ocean environment oxidize the nitrogen in ammonia to form the less toxic nitrite ion (NO_2^-). As long as these bacteria continue their activities, the free ammonia in the environment is held in check.
3. The nitrogen in nitrite is further oxidized largely by the action of another type of nitrifying bacteria (genus *Nitrobacter*), forming the even less toxic nitrate ion (NO_3^-). Thus, levels of nitrite are also kept in check.
4. Nitrates are used by photosynthetic organisms (marine algae) and, as a result, the nitrogen is used by the plant to make proteins or converted to the pure form, N_2 (aq), which is nontoxic.
5. Nitrogen can begin the pathway all over again via either:
   a) consumption and excretion by fish and other organisms
   b) conversion from nitrogen to ammonia via "nitrogen fixation" by microorganisms such as cyanobacteria

This is a simplified version of a complex biochemical cycling system that actually consists of a variety of additional interdependent processes. However, this scheme provides the beginning learner with a sound introduction to the concept of ecological chemical cycling in general and to the nitrogen cycle in particular.

Many of the major events of the nitrogen cycle can be elegantly demonstrated in the laboratory by the students themselves, using aquaria.

Procedure

1. Set up four small glass or plastic aquaria (4 to 10 L) using one of the many commercially available buffered seawater mixes. An undergravel or box filter powered by an air pump will be adequate to provide aeration, mechanical filtration and a substrate on which the aerobic nitrifying bacteria can establish themselves. A stable temperature (around 25° C) and eight hours of good lighting are required. Label the tanks A through D.
2. Tank A will serve as a control, so all conditions described above must be met (i.e., filtration, temperature), but no fish will be added.
3. Put one, two and three equal-sized saltwater-acclimated adult molly fish (genus *Poecilia*) into tanks B, C and D, respectively. Bacteria is introduced to the water in the slime coat and gut of the fish. Feed the fish flake food twice daily, the ration for each fish to be exactly the same.
4. Beginning with day 1, measure water from each tank with packaged test kits for ammonia, nitrite and nitrate. Based on the units of concentration used in your test kits, construct a graph (x-axis = time [in days]; y-axis = concentration [in units]) for each substance. Record all test results as points on the graphs. Note also the presence or absence of algae growing in the aquarium.
5. Continue maintaining identical conditions in all tanks (as much as possible). Measure the ammonia, nitrite and nitrate levels daily (or less often, to conserve test chemicals).
6. The expected result (the hypothesis that is being tested) is that ammonia should increase rapidly in the three experimental tanks until there is enough to sustain a significant colony of *Nitrosomonas*. As the bacteria multiply, the ammonia level should drop off significantly as the nitrite level increases (ammonia is being converted to nitrite). The nitrite level should rise for a time until the *Nitrobacte*r can establish themselves, at which time the nitrite levels should drop and the levels of nitrate begin rising. As the nitrates rise, algae will start to appear in quantity and begin using the nitrates. (Remember, there is a constant influx of nitrogen in the form of proteins in the fish food.) One would expect no changes in tank A, and varying rates of change in tanks B, C and D.

Notes

Ammonia toxicity may occur, especially in tank D, so be prepared to abort a test tank to prevent discomfort to the fish. A bacterial bloom at a peak in the cycle may cause the water to look cloudy and oxygen to be depleted rapidly (fish will be seen gulping air at the surface). Be sure to maintain the water at a constant level and specific gravity (1.020).

The time over which the cycling occurs will vary from tank to tank. The length of the experiment also depends on the size and health of the fish, the size of the tanks, rate of aeration, the temperature and a variety of other factors. The cycle can be rushed by inoculating the tank with bacteria from an established saltwater aquarium or with commercially distributed cultures (Patton 1987). Using the procedure described as is, it is best to allow a month for data collection, but two to three weeks often will do nicely. Instruct learners to stop data collection at a predetermined point (or ask them to determine the end-point themselves).

The test procedure provides a clear picture of some of the major events of the nitrogen cycle as colonization of a "new" marine habitat begins. It has the practical value of providing learners with the opportunity to use several important experimental methods: long-term data collection, use of experimental controls, use of animals in test situations, quantitative chemical analysis, graphing of results, interpretation of graphs and ecological modeling in laboratory environments. If students become interested in marine biology and begin contemplating saltwater aquarium as a hobby, the results obtained in this experiment can provide the basis for planning a properly established aquarium community.

Sources

All the materials described in this chapter can be obtained inexpensively from your local pet/aquarium supplier or any of the major biological supply catalogs. You may want to shop around for the best buy on saltwater test kits (in terms of both price and ease of use). Avoid using kits with toxic reagents and always use adequate safety precautions.

References

Dow, S. (1986). Incremental stocking. *Pet Dealer, 35*, 8-9.

Huckstedt, G. (1973). *Water chemistry for advanced aquarists.* Jersey City: TFH Publications.

Miller, G.T. (1979). *Living in the environment.* Belmont, CA: Wadsworth.

Patton, K.T. (1987). The nitrogen cycle in artificial marine habitats. *Marine Fish Monthly, 1*(12), 20-48.

Riehl, R. & Baensch, H.A. (1986). *Aquarium atlas.* Melle, W. Germany: Baensch (Tetra Books in U.S.).

Spotte, S. (1973). *Marine aquarium keeping.* New York: John Wiley and Sons.

Tortora, G., Funke, B. & Case, C. (1982). *Microbiology: An introduction.* Menlo Park, CA: The Benjamin/Cummings Publishing Co.

Can Beach Erosion Be Stopped?

Gregory J. Conway

As long as man has relied on ships for trade, people have settled along the coast. Most of the early villages were located along harbors carved out by nature. This afforded them protection from the immediate power of the sea.

Today, seven of the largest cities in the U.S. are on the coast and about 40 percent of our manufacturing occurs in coastal states. Studies show that 52 percent of the population lives within 50 miles of an ocean or one of the Great Lakes. Coastal real estate value is soaring, and, despite problems with pollution, the shore is becoming more and more popular.

But our shoreline is eroding away, especially along the Atlantic coast. In some places–North Carolina, for example–this has happened at rates of up to six feet a year. This could lead to disaster for those living near the shore, so we try to reclaim what has been lost; we try to stablilize our beaches. (Keep in mind that many think of a disaster only in terms of how it affects humans.)

Stabilizing Our Beaches

According to engineers, there are two approaches toward stabilization:

1. build a wall (groin) extending into the water perpendicular to the shoreline
2. build a wall parallel to the shoreline (seawall)

Groins are supposed to trap sand and therefore stop what is called its littoral transport along the coast. This movement is due to a current caused by waves approaching the beach at an angle. A jetty, cousin to the groin, is essentially a wall constructed on each side of an inlet. Jetties are supposed to prevent silt from building up in a channel and closing it off. Seawalls, on the other hand, don't really protect beaches, they protect the land behind the wall.

Engineers were called in because it seemed logical that if a town were losing sand, strategically placed obstructions would prevent any further loss. However, when a town constructs groins it cuts off the supply of sand to the city down current. That city, in turn, loses sand but no longer gets a new supply. It has no alternative but to construct its own groins. The result seems to be a domino effect.

Activity

Why is it so difficult to stop beach erosion, and why are manmade constructions considered only temporary? You can come up with your own answers by setting up a a small-scale model and studying the problem of erosion (see Figure 1).

Materials

- A stream table, which can be purchased through a supply company, or a watertight container roughly 150 x 60 x 10 cm
- 3 to 5 gallons of sand
- 30 to 50 rocks (averaging 5 cm in diameter)
- 2 bricks (for seawall)
- a strip of plastic or wood at least 10 cm high x 45 cm long
- a plastic ruler.

Procedure

Place the stream table in an area where it can remain undisturbed for a couple of days.

1. To study longshore current & the use of groins:

a) Add sand along one side of the stream table until it is about four or five inches wide, then add water gently, so as not to destroy your beach. Next, construct a series of groins with the rocks as shown in Diagram A of Figure 1. **Note:** The water level should be lower than the height of the beach and the tops of the groins.

b) Make small, consistently spaced waves by pushing the plastic strip back and forth for 10 to 15 minutes.

c) Observe the movement of sand.

d) At the end of the simulation, make sketches to show where sand accumulates and where it erodes.

2. To study longshore current and the use of jetties:

e) construct the beach and jetty simulating one side of an inlet as shown in Diagram B. Make sure the top of the jetty is above the water level.

f) Repeat procedures b, c and d.

3. To study wave action on beaches with a seawall:

g) construct the beach and brick seawall completely across the width of the stream table as shown in Diagram C.

h) place a ruler in the sand in front of the seawall to measure the beach height; record this as Point A. Then measure the beach height at the water's edge (Point B); this gives you the beach slope, as shown in Diagram C, side view.

i) Repeat procedures b, c and d.

j) Measure beach height at Points A and B, and compare to original measurements (step h). This can be seen more clearly if a graph is made.

When you are finished with the three models it may be

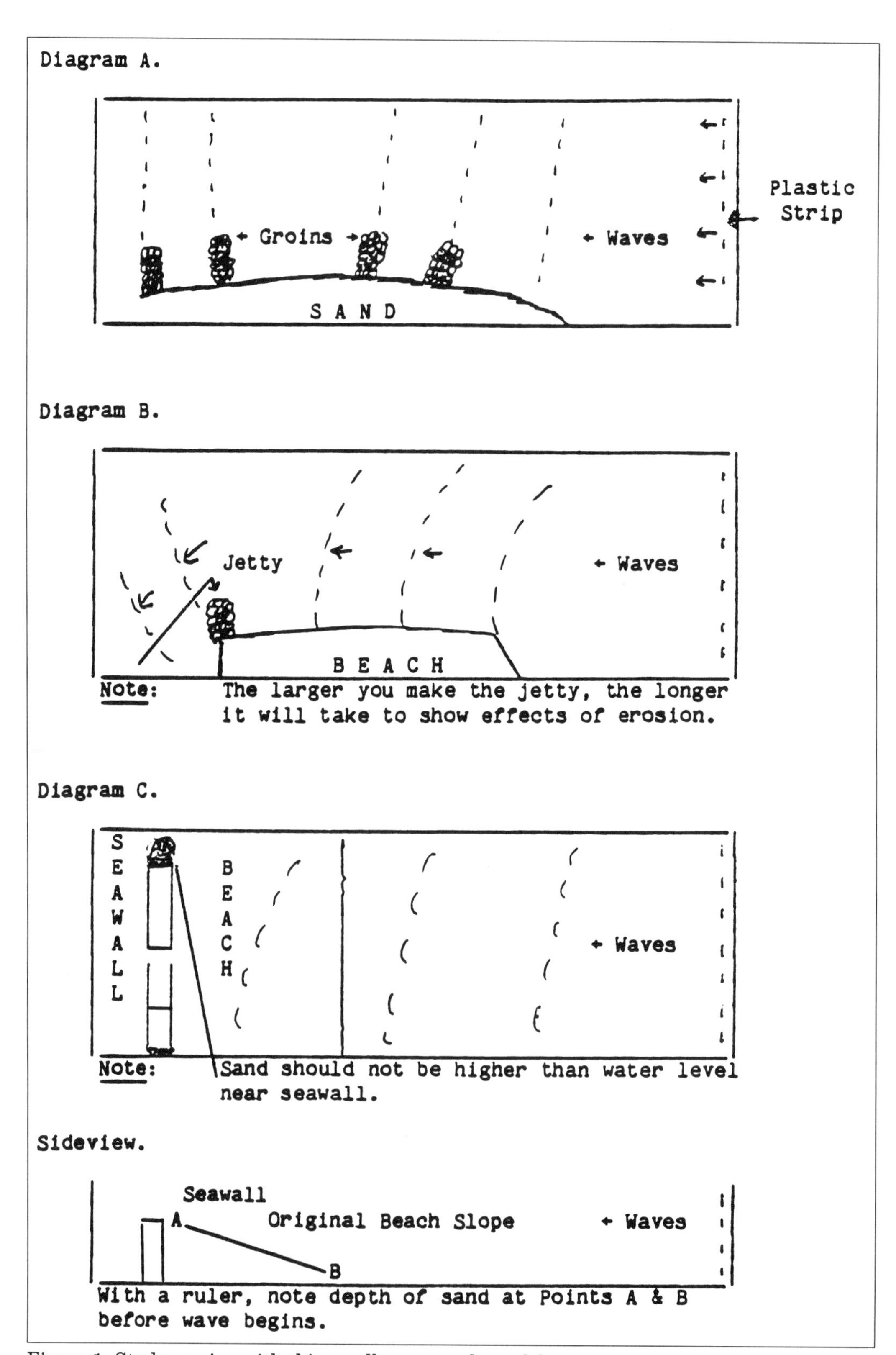

Figure 1. Study erosion with this small, man-made model.

apparent that such constructions could actually worsen the problem of erosion and should only be used as a last resort.

Erosion

Some scientists believe that stabilization is irreversible, that towns using manmade constructions are throwing their beaches further out of equilibrium. A beach that is out of equilibrium is no longer flat–it is steeper–and this change in profile influences how waves break on a beach. You can see from the seawall simulation that after a while sand is carried away from the construction and deposited further offshore. This means that there is nothing to slow down the waves as they approach the seawall. There is no friction to drain them of their energy. Consequently, they'll continue to hit the seawall full force with no other place to go but down into the sand, gouging it out further. Eventually, this process will undermine the construction which, without further intervention, will fall into the ocean. It seems like a problem with no solution, but that's only half the story.

The main cause of erosion is rising sea level. This is a controversial issue tied to glacier melting or inter-glacial periods between ice ages, the greenhouse effect and the depletion of the earth's ozone layer. Orrin Pilkey, a geologist, estimates a sea level rise of 12 to 15 inches per century. But this figure does not pertain to all locations. Local sea level changes can only be realized after years of recording and studying daily tidal fluctuations.

If sea level is rising at the same time that our beaches are being eroded by wave action and littoral currents, we must recognize the fact and face the situation with aggressive and relevant research. This research can begin in the classroom.

References

Bascom, W. (1980). *Waves and beaches.* Garden City, NY: Anchor Books.

Gilbert, S. (1986, August). America washing away. *Discover*, pp. 28-35, 75, 78.

Kaufman, W. & Pilkey, O. (1983). *The beaches are moving.* Durham, NC: Duke University Press.

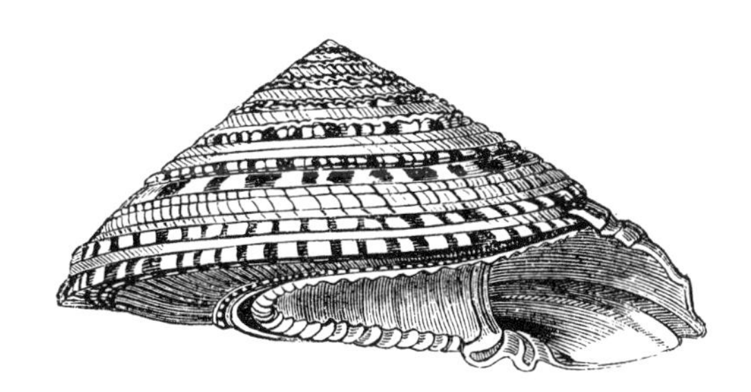

Simple Groundwater Investigations

James O'Connor

Groundwater leaks into streams and coastal wetlands to provide flow for perennial streams and tidal cycle fluxes.

Groundwater plays an important part in aquatic and human ecology. Subsurface moisture controls many facets of plant growth in soils and rock. Groundwater leaks into streams and coastal wetlands to provide flow for perennial streams and tidal cycle fluxes. Despite being underground and out of sight, groundwater concepts and processes can be easily modeled with homemade equipment and measurements. Student data should be compared to USDA-SCS Soil Survey tables. Federal soil surveys are available for most counties across the nation from the local extension agent.

Objective

Measure and understand five critical groundwater concepts for ecology.

Infiltration: ***The Cat Food Can Model***

How does rainwater or precipitation get into the ground? Infiltrameters are the instruments used to measure a storm's volume and rate of infiltration for different types of ground cover.

Materials

Remove the bottom of an empty cat food can so that you have a collar. You also need a water bucket and 100-500 ml plastic graduated cylinder.

Activity

1. Sink the collar halfway into the ground. (**Note:** You will have difficulty with grass and packed soil, but try.) Make sure the top of the infiltrameter is level. If water seeps out of the sides, cancel experiment data.
2. Pour a known volume of water into the collar and time the absorption of infiltration. Caution: Do not overfill. Use plastic 100 or 500 ml graduated cylinders for pouring water. Take a large bucket of water outside to draw from for your supply. Units are measured in ml/s compared to in/hr in soil surveys.
3. Measure infiltration rates outside on school grounds in three different environments, for example, mulched, sandy and grassy areas. Compare those rates with infiltration rates

you would predict for an asphalt or concrete area. This comparison will show the difference between infiltration and runoff in an asphalt surface or urbanized environment as compared to natural surfaces like farmland or forest.

Questions

If rain water does not become groundwater, what does it become? What happens when the groundwater freezes in the winter? What happens when permafrost thaws out? Does the temperature of the water make a difference in the infiltration rate? Which surface material has the highest infiltration rate?

Porosity: *The Sponge Model*

How much air space is in the soil or rock for water to occupy? Most lab investigations for measurements of porosity are done with gravel chips, sand, or different-sized beads placed in plastic tubes or graduated cylinders. While a set volume of sediments (e.g., 100 ml) has a bulk volume, you must measure the air space in soils and sediments of this volume. Porosity is measured in percent (air space volume over the total or bulk volume x 100).

Materials

sponge plus 50 ml of water in a graduated cylinder

Activity

1. Lay the sponge flat.
2. Measure the volume (1 x w x h) of the dry sponge and record it.
3. Slowly and carefully pour the water into the middle of the sponge. Keep pouring in water until it leaks out of the sponge or will not hold any more.
4. Record the volume (ml) left in the graduated cylinder.
5. How much water is in the sponge?
6. Remeasure the volume (1 x w x h) of the wet sponge. Explain

Selected values of porosity, specific yield and specific retention

[Values in percent by volume]

Material	Porosity	Specific Yield	Specific Retention
Soil	55	40	15
Clay	50	2	48
Sand	25	22	3
Gravel	20	19	1
Limestone	20	18	2
Sandstone (semiconsolidated)	11	6	5
Granite	1	.09	.01
Basalt (young)	11	8	3

the difference.

7. Squeeze the sponge in a beaker and pour this water back into the graduated cylinder. If you did not spill any, why did you not get all the water back? Look up "capillary action" and "specific retention" as special groundwater vocabulary words.
8. Look up the porosity of local soils as listed in the local soil survey.
9. Compare porosity for different soil types or sediment sizes.

Questions

How does porosity control all underground liquids or gases–e.g., oil, gas, saltwater, leachate and radon–during migration or ponding? How are porosity and infiltration related to the difference between a swamp and a marsh? How does detritus get out of a marsh, and what is left behind? What are normal porosity values for different sediments, soils, or rocks? If a liquid rises in the ground, what will happen to any lighter material higher than the rising liquid? What does compaction do to porosity?

Permeability:

The Flower Pot Model

Can water go through the ground? If so, at what rate? As with porosity, plastic tubes that are open at both ends and filled with different sediments, soils, or sizes of beads have different rates of flow or drainage through the geology or soil (sieve effect). Most geology has secondary permeability because of cracks (joints or fractures). Permeability rates calculate out as a distance for a unit time ($v = d/t$). Percolation is movement within the body. Transmissivity is movement horizontally from the body.

Materials

Obtain several similar flower pots and fill with different soils. Plastic tubes and soils used in porosity experiments can be reused.

Activity

1. Read the steps first.
2. Pour a set amount of water from a graduated cylinder (100 ml recommended) into the soil.
3. Capture the water from the bottom of the pot in a beaker and measure the amount.
4. Time the process until the steady drips stop.
5. Record the data in a table (water in, water out, time).
6. Repeat the process with the same plant pot three times. Replenish the 100 ml supply each time.
7. Test the chemistry of the water. Compare the water going into the pot to the water coming out for pH, temperature, color, or other characteristics.

8. Try other plant pots with different soil mixtures. If available, use marsh and swamp soils, or try a mixture of pure peat, sand, or marble chips.
9. Record and compare each experiment.

Note: Make sure all pots are the same size. Multiples of 100 ml of water are recommended to simplify calculations.

Questions

What does any ground excavation do to permeability? What does pumping do to permeability rates? How do flow rates relate to organisms? What will happen to flow rate at a tilted boundary between a permeable and impermeable geology or soil horizon? What happens to leachate or pollution as it seeps or permeates to the next layer? Which way do liquids move below the surface? Can you use this moving groundwater energy for a resource? How would you classify groundwater seepage from tidal marshes or coastal landforms during ebb tides?

Artesian Concept: ***The Plastic Bottle Model***

How does an artesian well work? There are two kinds of groundwater systems: normal and abnormal. A normal system is a ground watershed related to the local stream where there is direct influent and effluent seepage during the seasons. An abnormal groundwater system is a confined aquifer system where the water is under confined pressure. This means it is sandwiched between the overlying and underlying rock units. Groundwater moves mainly by gravity and hydrostatic pressure and is a function of topography and "swiss cheese" geology.

Materials

Two plastic squirt bottles connected by a tube. Place a pin hole in the bottom of each bottle and connect the tube to the squirt tops. Fill each bottle three-fourths full with regular water.

Activity

1. Turn the bottles upside down and mark the water table in each. Observe the zone of aeration and zone of saturation.
2. Move one bottle higher than the other and watch what happens.
3. Time the change for 30 seconds and mark the water table changes in each bottle. (Don't move the bottles but have two students ready with a marker for each one.)
4. Wait for the artesian spring to develop in the lower bottle. When you have a spring, raise the higher bottle higher.
5. Lower the higher bottle to an equal water height with the other bottle. Do it gently, not quickly.
6. Now reverse the process with the other bottle being raised higher.

Note: Watch the water loss in each bottle. Replenish the water it you lose too much or the spring does not develop quickly in the other bottle.

Questions

How can fresh water springs occur under the ocean on the continental shelf? What does confinement do to water pressure? How does water move inside a tree? Why are many artesian systems seasonal? If water pressure is greater than air pressure, what does the water table represent? What happens when the water tables are at an even elevation? How is groundwater energy used by man? Why does salty groundwater replace fresh groundwater, especially in coastal and estuarine environments?

Groundwater Hazards: *The Soup Can Model*

Groundwater changes that occur from the addition or subtraction of water may cause different types of gravity transfer, such as landslides, mudflows, or sinkholes. Groundwater has mass and volume. Too much infiltration, especially on sea cliffs, or dewatering by overpumping may cause life and property-threatening changes, as well as new landforms for microenvironments or ecological habitats.

Materials for homemade landslide kit

- 2 empty tin soup cans
- a rectangular, plexiglass sheet.

Remove the brand labels from cans and label one A, the other B. Can B is best used as the permeability can. The bottom of can B needs to have 15 to 20 ice pick holes (small), while can A has none. This is a test for permeability.

Note: Both cans have 100 percent porosity.
Oral Test: Ask students which can has porosity: can A or B. Obtain a clear plexiglass sheet about one foot long and one-half foot wide (a cafeteria tray or wood sheet is fine). Fill two 100 ml graduated cylinders with water. Obtain or make a protractor.

Activity

1. Practice and test the angle of repose or slide angle for the dry cans. Set the two cans at one end of the plastic sheet and lift this end until sliding occurs.
2. Measure the angle of slide with the protractor laid next to the plastic sheet. The sheet should be raised gently but constantly until the cans slide. This next activity requires five students: one for lifting, one for holding and measuring the angle of repose at the other end of the sheet, one as a rainmaker for can A, one as a rainmaker for can B and one as a recorder. Repeat each experiment three times and record the average angle of repose between the three.
3. Before using the water, have students make hypotheses

about which soup can (A or B) will slide downhill first after infiltration.

4. Which cans will go lower or higher than the angle value for the dry condition and why?
5. Test the rain hypothesis. Record and repeat the experiment three times. Discuss the results.
6. Have students experiment with other conditions, such as a wet sheet, etc. Record all data in a table.
7. Have students develop one table for all experiments on each can and condition.
8. Have students look up where and when real examples of each of their experiments might occur locally, nationally, or globally.

Questions

Relate this investigation to movement on dry and wet sand dunes. Relate this activity to slides and slumps from shoreline cliffs of Great Lakes, California, northern New Jersey and Chesapeake Bay's western shore. How does the weight of water relate to landslides? How does gravity transfer relate to the boundary between two different geologies like cans and sheets? What kind of landforms result from sliding along shorelines? Why do the rainy season and the landslide season coincide for many regions? Why are groundwater-produced slides ideal sites for the preservation of fossils? Do submarine slides differ from terrestrial slides in mechanics or causes?

References

American Institute of Professional Geologists. (1983). *Ground water issues & answers.* Arvado, CO: Author.

Baldwin, H.L. (1963). *A primer on ground water.* U.S. Geological Survey.

Boyd, S. et al. (Eds.). (1984). *Groundwater: A community action guide.* Washington, DC: Concern.

La Fleur, R.G. (Ed.). (1984). Groundwater as a geomorphic agent. *Binghamton symposium in geomorphology #13.* Boston: Allen & Unwin.

Geophysics Study Committee, National Research Council. (1984). Groundwater contamination. *Studies in Geophysics.* Washington, DC: National Academy Press.

Heath, R.C. (1983). *Basic ground-water hydrology.* (U.S. Geological Survey Water Supply Paper 2220). Washington, DC: Government Printing Office.

Heath, R.C. (1984). *Ground-water regions of the U.S.* (U. S. Geological Survey Water Supply Paper 2242). Washington, DC: Government Printing Office.

O'Connor, J.V. (1985, Fall). Chesapeake Bay's underworld comes to light. *Current, Journal of Marine Education, 6*(4), 18-23.

Price, M. (1985). *Introducing groundwater.* Boston: Allen & Unwin.

U.S. Geological Survey. (1985). *National water summary—1984.* (Water Supply Paper 2275). Washington, DC: Government Printing Office.

U.S. Geological Survey. (1988). *National water summary—1986: Ground water quality.* (Water Supply Paper 2325). Washington, DC: Government Printing Office.

Divide & Conquer: The Story of a Watershed

Vicki Price Clark
Teresa Auldridge

The cycling of water through our physical environment connects all of us to the ocean, no matter how far inland we live. As water journeys downstream through our communities, pollutants and sediments are added as we take showers, wash our cars, water our lawns and fertilize our gardens. Heavy sediment loads and the accumulation of toxic materials from industry, agriculture and residential areas are major forms of pollution damaging our lakes, bays and oceans.

To appreciate how inland activities can impact water quality miles away, students need to understand the concept of a drainage basin, or watershed. A watershed is an area of land that is drained by a stream or a river and all its tributaries. A watershed may be small, such as the area on the sides and head of a gully, or it may be a large and complex system of streams and creeks which drain into one major river.

Watersheds are separated from one another by high ridges or "divides." The Continental Divide of the United States, for example, is in the Rocky Mountains. All the precipitation falling on the west side of the divide flows into the Pacific Ocean; all the precipitation falling on the east side eventually flows into the Atlantic Ocean. The largest single watershed in the United States is that of the Mississippi River which includes all the land between the Rocky Mountains and the Appalachian Mountains.

Pollution

Pollutants introduced into a watershed can affect the water quality downstream. Human activity creates both point source and non-point source pollution. Point source pollution comes from a specific, easily identifiable place, such as a discharge pipe from a factory or sewage treatment plant. Non-point source pollution results from the runoff of substances such as fertilizer, automobile oil and pesticides from lawns, parking lots and streets into storm sewers, which then flows into creeks and rivers. Other non-point sources of pollution are soil and chemical erosion from forests, farmland and construction sites. All these pollutants can have a negative impact on aquatic organisms in rivers and in the ocean.

The following activity has been designed to help students learn the meaning of the terms "watershed" and "divide," and provide them with a simulated experience of how water moves through a drainage basin. Using simple materials the students will create their own miniature watershed and observe the movements of precipitation and pollutants. This activity has been used successfully with students from the fourth grade through high school, as well as with adult groups.

Materials

(for each group of four to six students)

- shallow, waterproof box, such as a plastic storage box or an aluminum roasting pan
- aluminum foil
- sprinkling can or similar device (a cup with small holes punched in the bottom will work)
- water
- food coloring
- cotton balls
- white glue
- topographic map of local watershed (available from a state geologist or local conservation agency)

Procedure

Use the shallow box as the base container for the model. Cut a piece of foil about one and one-half times the size of the box. Create an artificial landscape by molding the foil into hills, ridges and river valleys, tucking the edges of the foil inside the box. Predict the direction of water flow over the model. Next, "rain" on the landscape by sprinkling water over it. Observe the direction of the water flow and identify the divides and watersheds in the landscape.

What effect would a toxic pollutant have on this area? Simulate a toxic spill by carefully placing a drop of food coloring in one place on the foil. Now sprinkle water over the foil again and observe where the color moves. What parts of the landscape are affected? Discuss why the clean-up of such spills is so difficult.

How is this artificial landscape different from a real land surface? (Without soil and vegetation, all the precipitation runs off into the rivers and streams.) Using cotton balls and some white glue, "plant" vegetation in various places in the landscape. Predict how water flow will differ from that in the first experiment. Repeat the toxic spill and observe where the coloring ends up.

Using a topographic map of a local river basin, identify the boundaries of its watershed. What small tributaries flow into the main stream? What major cities are located in the watershed? What kinds of pollutants (point source and non-point source) might be affecting the area?

Extension

Take the students out in the schoolyard and have them make inferences as to what the water flow patterns are when it rains. Give them sprinkling cans and let them simulate rainfall, identifying small divides around the yard. Have the students draw maps of the area indicating the water flow patterns.

This activity is one of 31 science and social studies exercises included in *River Times*, a curriculum developed at the Mathematics and Science Center in Richmond, Virginia, with a grant from the Virginia Environmental Endowment through Richmond Renaissance. *River Times* focuses on the natural and cultural history of the James River and Belle Isle.

References

Citizens Program for the Chesapeake Bay. (1985). *Baybook: A guide to reducing water pollution at home*. Baltimore, MD: Author.

Hamblin, W.K. (1982). *The earth's dynamic systems: A textbook in physical geology*. Minneapolis, MN: Burgess Publishing Co.

Mathematics and Science Center. (1987). *River Times*. Richmond, VA: Author.

Pipkin, B.W. & Cummings, D. (1983). *Environmental geology: Practical exercises*. Belmont, CA: Star Publishing Co.

United States Department of Agricultural, Soil Conservation Service. (1985). *What Is a Watershed?* Washington, DC: U.S. Government Printing Office.

Celestial Oceanography: Understanding Tides

J. Garrett Tomlinson

There exists a relationship between living organisms and the motion of celestial objects, particularly the earth and moon. All life on earth is influenced by the rhythms of the days, seasons and tides. This is the astronomical effect of a spinning planet which, accompanied by the moon, revolves around the sun. These rhythms of life determine the timing of biological activity and processes. One of the most important celestial influences on marine organisms is the fluctuation of tides. It is in the rhythm of the tides that astronomy and oceanography coalesce.

The significance of tides (the periodic rise and fall of sea level) is important to any study of oceanography. Humans depend on a knowledge of tides for navigation, fishing and coastal activities. Currents caused by the ebb and flow of tides generate a continuous circulation of sediments, detritus and nutrients. These erosional currents flow through channels, subsequently changing the shape of inlets and islands. The tides alternately submerge and expose land (the intertidal zone) creating a unique environment for plants and animals. There may be large fluctuations in salinity, temperature and oxygen level of the water as well as substrate. Organisms must be adapted to the daily environmental changes of the intertidal zones.

Moon Watching

To truly understand tides you must comprehend the relative positions of the moon, earth and sun during the monthly phasing of the moon. A good starting point for understanding this is a "moon watching" activity that requires students to observe the moon over a period of time. First determine the date of the new moon by using a calendar or almanac. One or two days after the new moon, direct students to look for the moon at sunset every other night and notice its shape and position in the sky relative to the sun. Students' observations should be recorded on a drawing similar to Figure 1 on which the date, familiar horizon objects and directions are labeled. The observations should be done over a period of approximately two weeks ending with the full moon. Each observation should be done from the same position at about the same time. By watching the moon over a period of time, shifts in its position and its changing phases will become apparent.

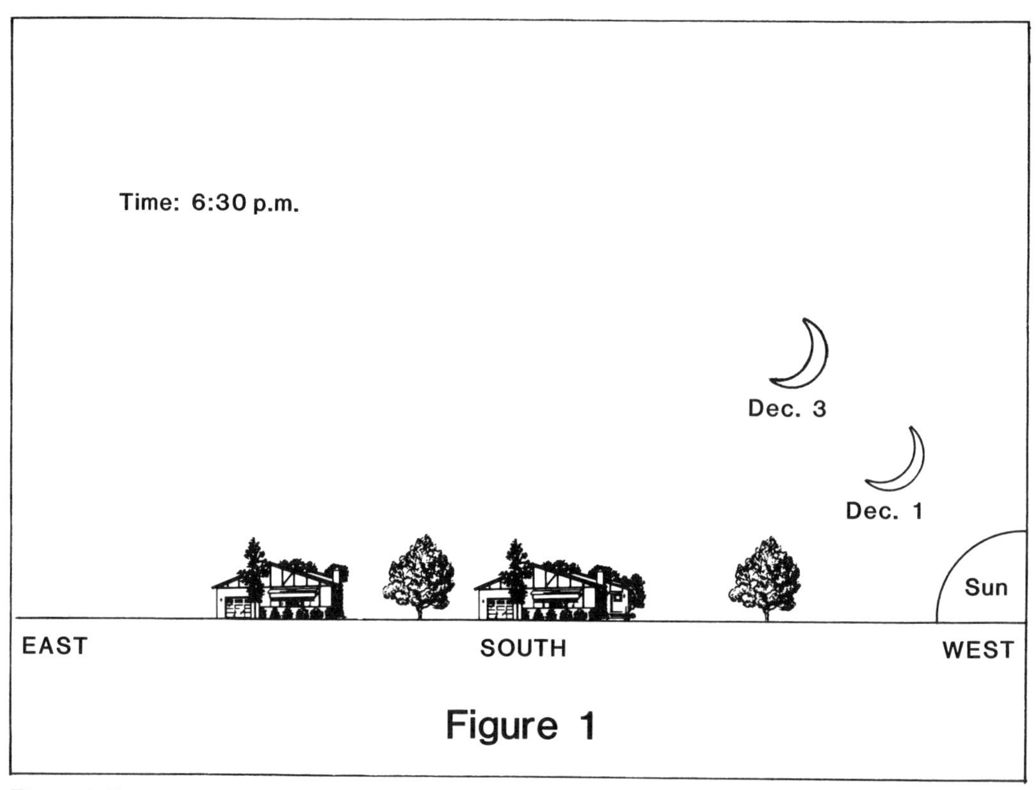

Figure 1. Example of drawing for student observations.

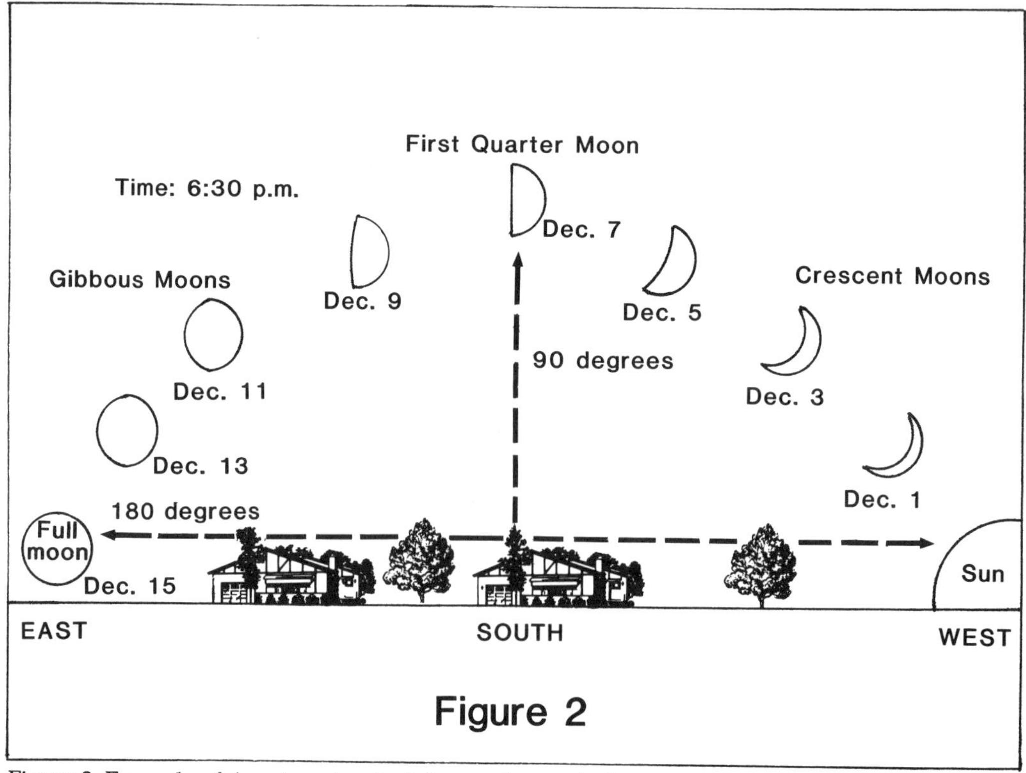

Figure 2. Example of drawing at end of observation period.

At the end of the observation period, the completed drawing should be similar to Figure 2. Examine students' drawings, noting the phase of the moon and its angle in relation to the earth and sun. Referring to Figure 2 as an example, note that crescent moons (Dec. 1-5) are less than 90 degrees of the moon-earth-sun angle while on December 7, the first quarter moon (half moon) is 90 degrees from the sun. The gibbous moons (Dec. 9-13) are between 90 degrees and 180 degrees from the sun. Note the full moon (Dec. 15) is opposite the sun at a 180-degree position.

Moon Phasing

After observing the positions of the moon, sun and earth in each moon phase, the concept of the moon phasing is easier to understand. The moon has phases because we see different portions of the lighted side of the moon as it orbits the earth. Since the sun supplies the light by which we see the moon, the apparent shape of the moon we view depends upon the moon's location in the sky relative to the sun. The sun always illuminates one-half of the moon (the inner circle of moons in Figure 3), but the half seen from earth usually does not coincide with the illuminated half. The half of the moon we see will contain none, part, or all of the illuminated side depending on the position of the moon in its orbit (the outer circle of the moons in Figure 3).

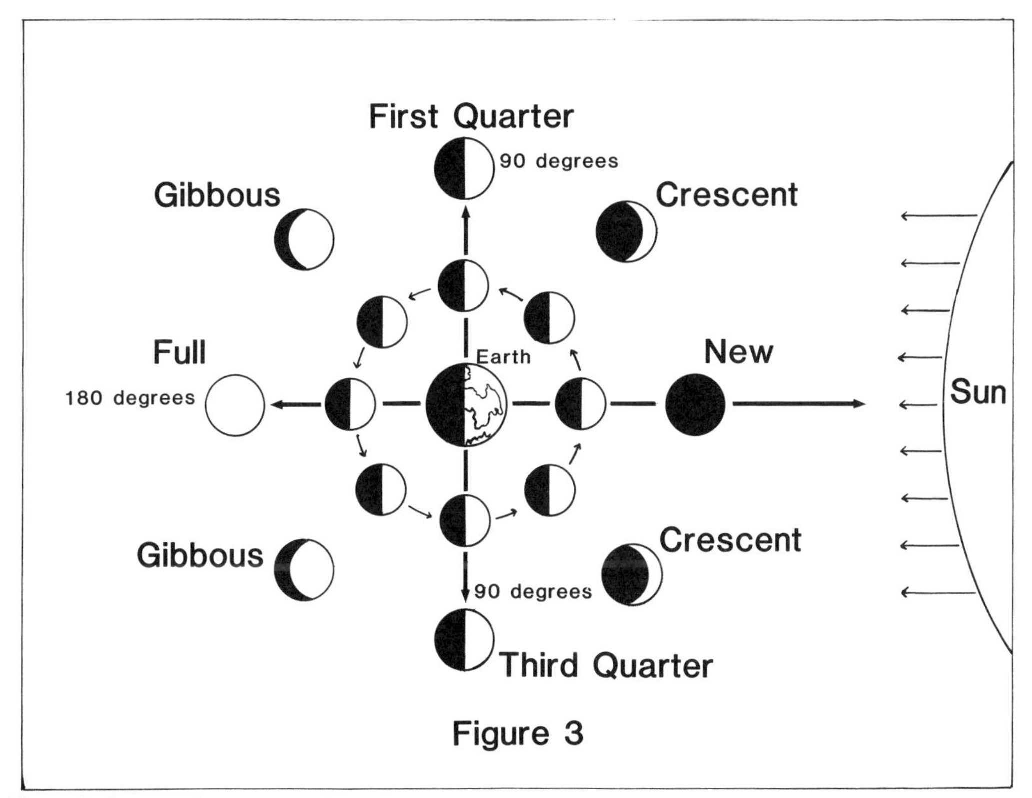

Figure 3. Sun–earth–moon angles.

When the moon is between the earth and the sun, the side toward the earth is dark (new moon). When the moon reaches exactly 90 degrees away from the sun, it appears half-dark and half-light (first or third quarter moon). Lying exactly opposite from the sun (180 degrees), the entire lighted side of the moon is viewed from the earth (full moon). Relate the sun-earth-moon angles seen in Figure 3 to the angles of the various phases noted during observations in Figure 2.

The best activity to help students visualize the phasing of the moon requires an overhead projector and a plastic foam ball (4 in. diameter) stuck on the end of a toothpick. Have a student stand about eight feet from the projector. Turn out the lights, and have the student hold the plastic foam ball at eye level (Figure 4). As the student slowly spins around in the light, the lighted portion of the ball will appear to change (Figure 5).

Gravity's Effects

Tides are caused by the gravitational attraction of the sun and moon on the earth's oceans. Seawater, being a liquid, is more easily influenced by gravity than the solid earth. Imagine the earth being uniformly covered with water with no land masses. The sun's gravity creates a bulge of water toward the sun, and the centrifugal force of the motion of the earth revolving around the sun causes a second bulge of water opposite the sun (Figure 6). Although the sun is much larger than the moon, it exerts tidal forces less than one-half as strong as the moon because of its greater distance from the earth.

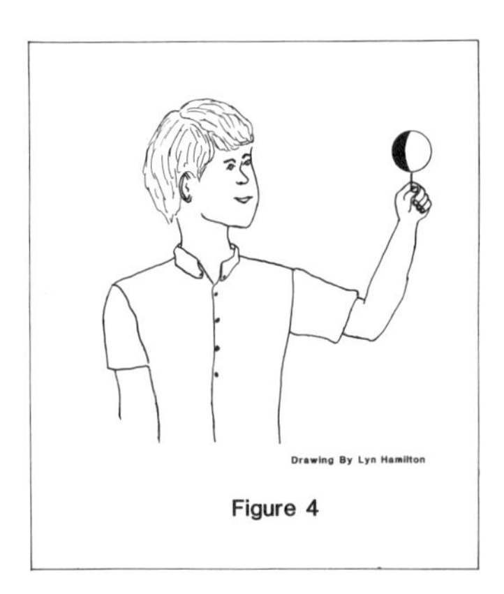

Figure 4. As part of an exercise to help students visualize the phasing of the moon, a student stands in front of a projector, holding a plastic foam ball at eye level, with the lights out.

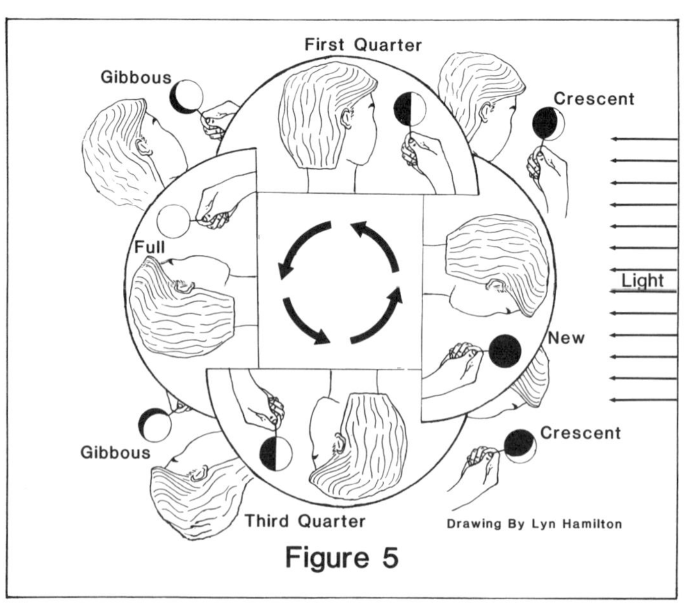

Figure 5. The student slowly turns around in the light from the projector. The lighted portion of the ball changes.

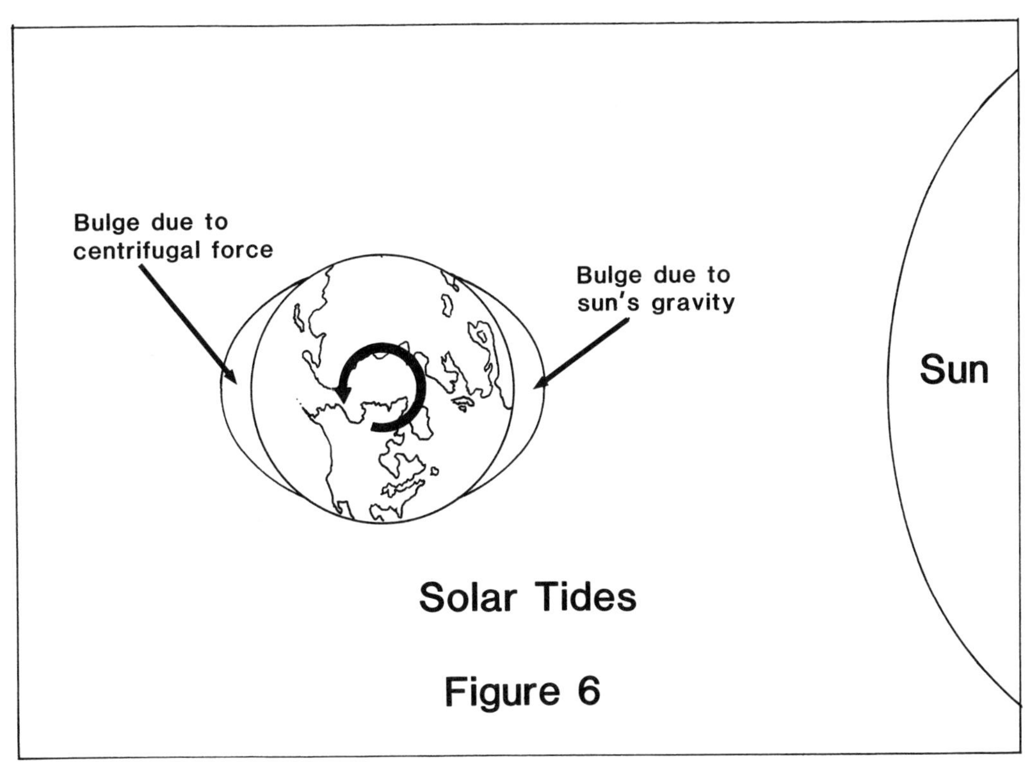

Figure 6. Solar tidal bulges.

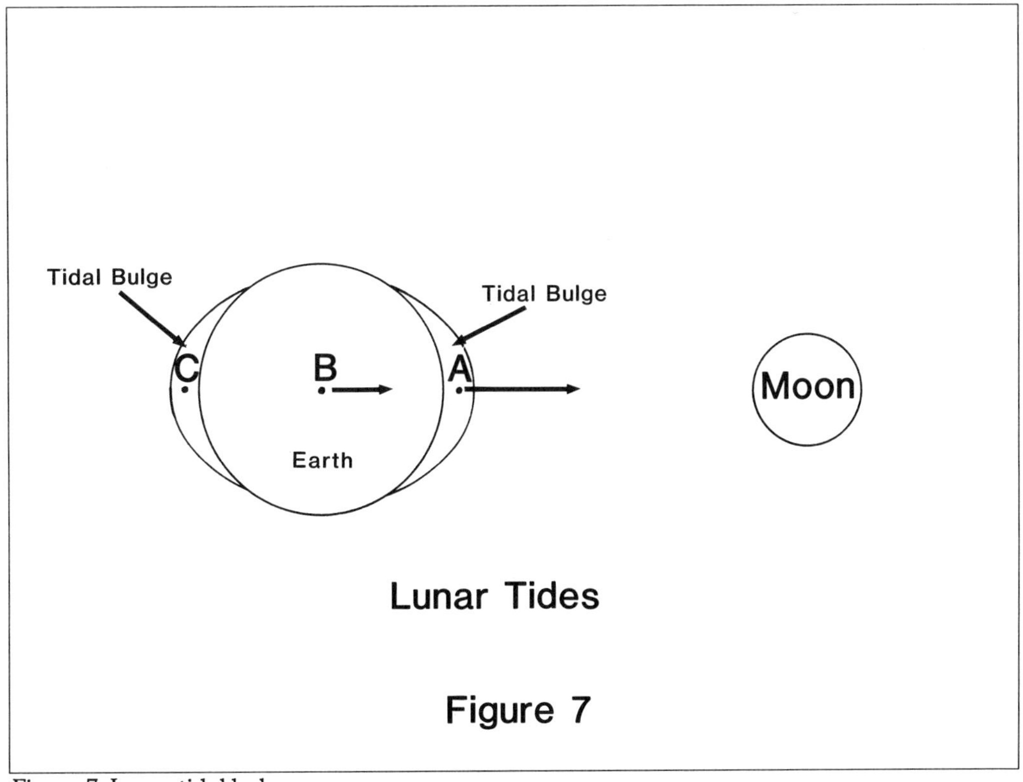

Figure 7. Lunar tidal bulges.

Tidal Bulges

The moon produces tidal bulges because of its differential gravitational pull across the earth's diameter and the centrifugal force of the earth-moon system. The earth-moon system revolves around a mutual center (barycenter) about 4,700 kilometers from the earth's center, creating a centrifugal force on the earth's waters, particularly opposite the moon. The moon attracts the water on the side of the earth facing the moon (Figure 7, point A) more than it attracts the earth, producing a tidal bulge in that direction. This is because gravitational force lessens with increasing distance, and the seawater is about 6,500 km closer to the moon than the center of the earth. Simultaneously, the earth's center (point B) is attracted by the moon with more force than the water on the opposite side of the earth (point C). Due to the earth being pulled away from the water on the far side and the centrifugal force of the earth- moon system, a second tidal bulge opposite the moon is produced. Because the moon exerts larger tidal forces than the sun, the tides usually follow the moon with its influence modified by the sun's relative position. The bulges remain fixed with respect to the moon as the earth rotates daily. When part of the earth sweeps beneath one of the bulges, high tides are experienced and areas in the depressions between the bulges encounter low tides.

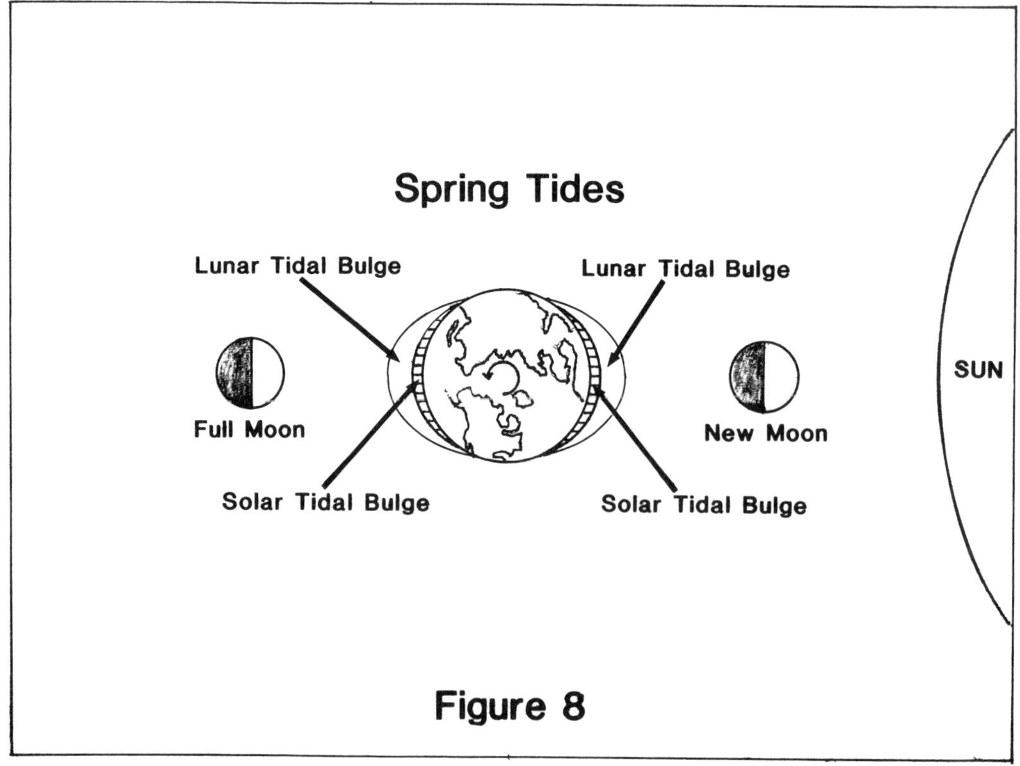

Figure 8. Spring tides occur during new and full moons.

High & Low Tides

Since the earth's rotation rate is 24 hours, the expected time between two high tides should be 12 hours. However, due to the moon's motion in its orbit, the earth has to rotate an extra 25 minutes to enter the next high tide bulge. Thus, the time between one high tide and the next is ideally 12 hours, 25 minutes. Realistically, the timing of tides and the amount of the tidal ranges can be influenced by land masses, the shape of the ocean basin, friction of water with the ocean bottom and the changing distances of the sun and moon from the earth. Also, due to the tilt of the earth and the inclination of the moon's orbit, the pairs of highs and lows normally will not be at the same level.

The interaction of the sun and moon in producing tides can be better understood using the information students have learned about moon phases. During new and full moons (sun-moon-earth and sun-earth-moon angle is 180 degrees) the tidal bulges of the sun and moon coincide (Figure 8) producing the highest and lowest tides, called **spring tides** (this refers to the way the water "springs" away from the earth rather than to the season). During the first and third quarter moons (the earth-moon-sun angle is 90 degrees), the tidal bulge of the moon is at right angles to the sun's tidal bulge, each partially canceling the other's effect

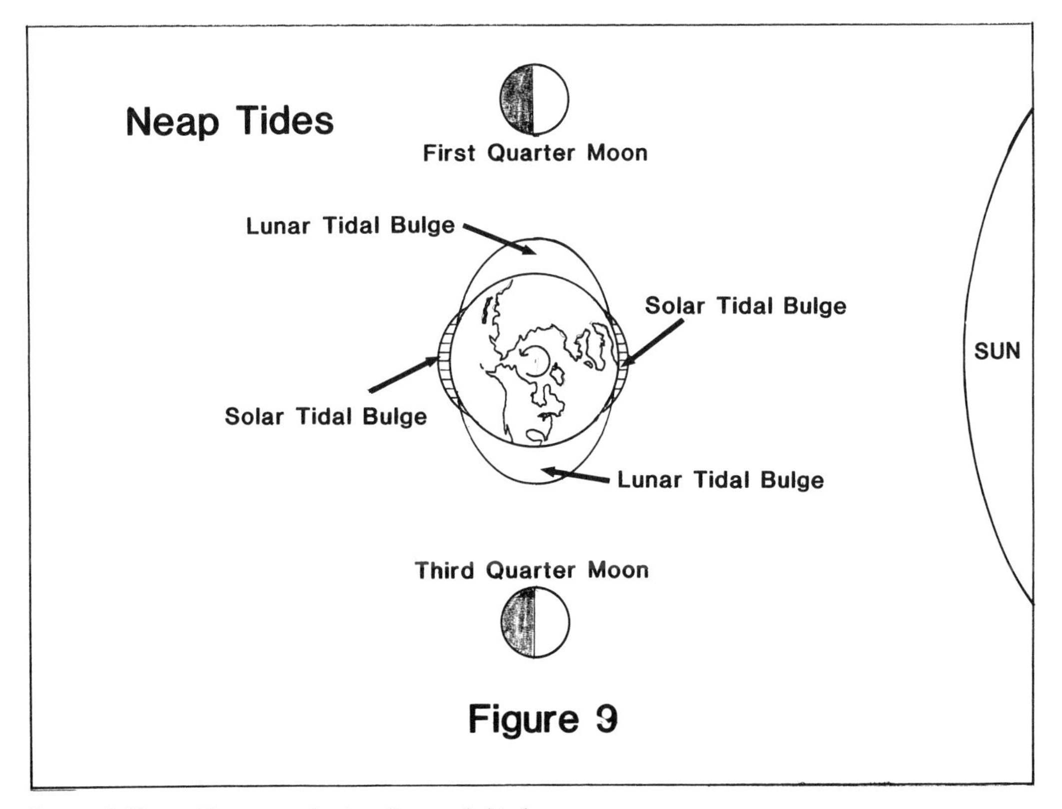

Figure 9. Neap tides occur during first and third quarter moons.

CELESTIAL OCEANOGRAPHY: UNDERSTANDING TIDES

Student Worksheet*

NAME____

PART A

Sketch in the tidal bulges on the earth for the alignment of the sun, moon, and earth below.

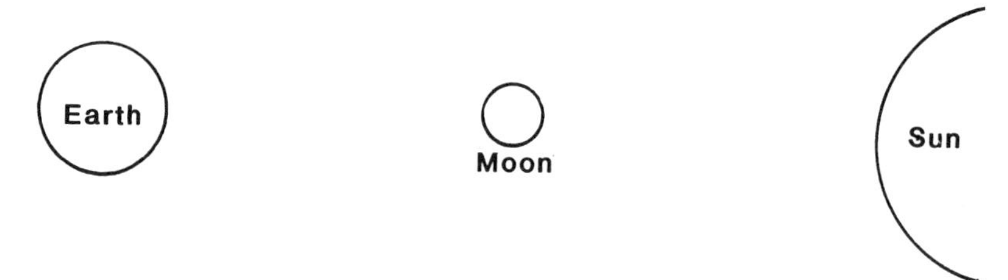

PART B

Draw in the moon in its correct orbital position for the phase given. Sketch the tidal bulges on the earth due to the effects of the sun and moon. Label the name of the tide.

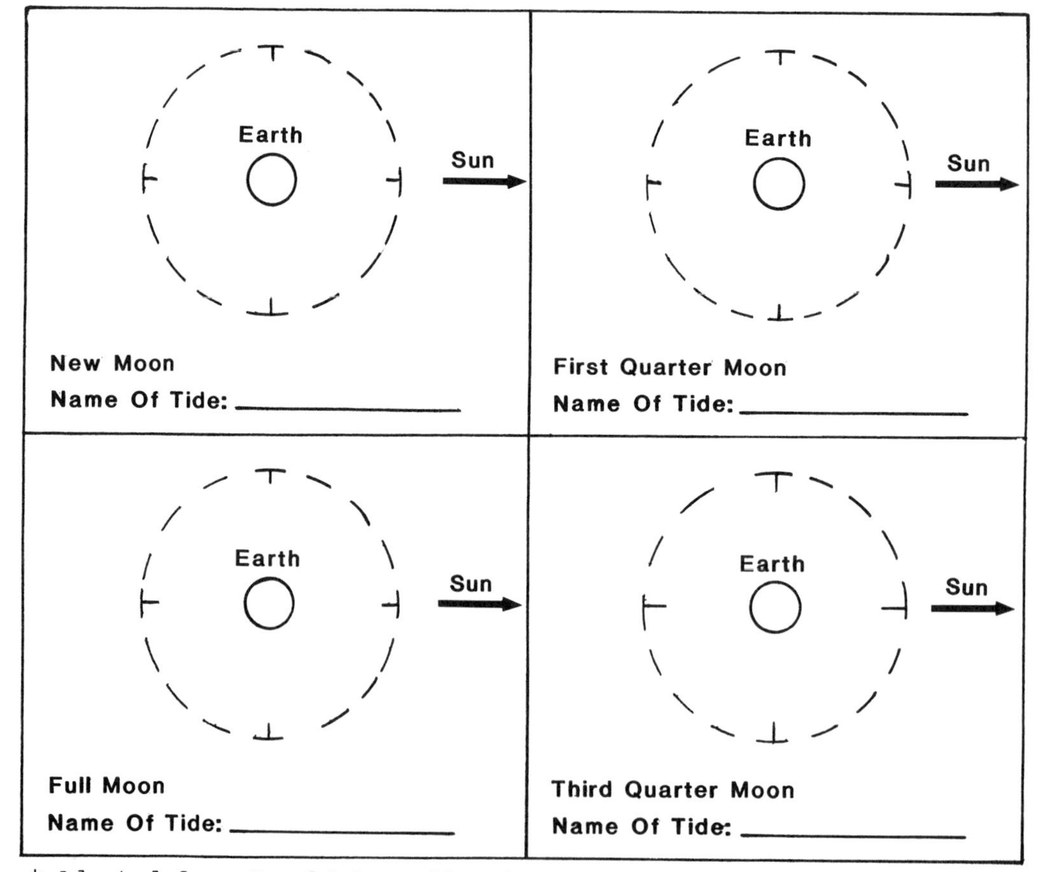

* Adapted from Gerald L. Mallon (1987).

(Figure 9). The result is the lowest high tides and highest low tides, called **neap tides**. "Neap" is a word meaning "hardly enough" in ancient Scandinavian language.

In Part A of the student worksheet, direct students to draw in the tidal bulges on the earth both toward and opposite the moon/sun. In Part B, students draw the moon in its correct orbital position for the indicated phase and sketch the tidal bulges on the earth naming the correct tide (spring or neap).

For an additional activity, have students use tide tables to determine the approximate dates of the four main phases of the moon (new, full, first and third quarter), noting when spring and neap tides occur. To secure tide tables of local coastal regions, write for the National Ocean Service (NOS) *Tide Tables* publication from the Superintendent of Documents, Government Printing Office, Washington, DC 20402.

In ancient times, it was thought that the sea and the sky were mirrors of each other. In the instance of tides, it is a mirror of relationships. Understanding tides can help students appreciate the connection between the sun-earth-moon system and the rhythmic behavior of marine organisms.

References

Carson, R. (1955). *The edge of the sea.* New York: The New American Library.

Davis, R.A., Jr. (1977). *Principles of oceanography.* Reading, MA: Addison-Wesley Publishing Co.

Dixon, R.T. (1984). *Dynamic astronomy.* Englewood Cliffs, NJ: Prentice Hall.

Jefferys, W.S. & Robbins, R.R. (1981). *Discovering astronomy.* New York: John Wiley & Sons.

Gross, M.G. (1967). *Oceanography.* Columbus, OH: Charles E. Merrill Books.

Mallon, G.L. (1987). Bioclocks: Understanding the rhythms of life. *The Planetarian, 16*(2), 36-48.

Maloney, E.S. (1985). *Chapman piloting, seamanship & small boat handling.* New York: Hearst Marine Books.

Mauldin, L. & Frankenberg, D. (1978). *North Carolina marine education manual: Seawater.* Raleigh, NC: UNC Sea Grant Program.

Spitsbergen, J.M. (1980). *Seacoast life: An ecological guide to natural seashore communities in North Carolina.* Chapel Hill, NC: University of North Carolina Press.

The Electronic Ocean

L. W. McLamb

So, you want to teach marine science in Kansas? A field trip to the ocean may be out of the question, but the resourceful teacher can find ways to bring the ocean into the classroom electronically, using microcomputers and software. Times are changing! Oceanographers today use airplanes and lasers to study ocean currents and track tagged sea turtles with satellites. The computer is not only an effective teaching tool; it is a scientist's tool and "the science and technology components we teach must be updated to reflect recent advances in these areas" (Klemm 1988). It is my hope that, after reading this chapter, fish and "chips" will take on a new meaning for many of you.

Bringing your classroom into the computer age takes only two ingredients: computers and software. The type of hardware and the number of computers available to teachers can differ greatly from school to school and is a factor the classroom teacher often cannot control. Teachers, however, have been successful in integrating computers into the science curriculum whether they have a lab full of machines or access to just a single computer on an occasional basis. Having access to good computer programs is the most important element in enhancing science learning using computers, and teachers can influence the software selected for use in their schools.

Tutorial Software

Computers can be used to present information on a number of marine topics. Educational Images produces a series of programs that discuss whales, sharks, coral reefs and ocean currents. The programs use good graphics and some sound effects to maintain student interest. Each program contains an interactive portion that is usually in a question and answer format. This type of approach is typical of the tutorial format, and the instructional advantages provided include self-pacing by the learner, immediate feedback in the question/answer sections, responses to student input that are non-judgmental and the opportunity for easy review of material introduced. Most important, the computer is a patient tutor.

TYC Software produces a program called *Shore Features* that teaches beach geography, and M.E.C.C. has a program called

Ducks that incorporates videogame features to teach waterfowl identification. A tutorial program with a twist is *Sea Life*, produced by Spectrum Software. This program comes with a sea life kit containing shells, starfish and urchins. Students, directed by the computer, make observations about these marine remains.

Tutorial programs should be thoroughly previewed by the instructor to determine if the information presented is appropriate for the intended audience. Tutorial programs are much like a textbook on computer; using the computer may help motivate the learner.

Simulation Software

Another category of software is the simulation program, which models natural occurrences through open-ended exploration of real-life situations. This uses the computer's unique capabilities more effectively than tutorial software.

A good example of a simple simulation is the M.E.C.C. program *Odell Lake*. In it, a student plays the role of a fish and encounters other fish or mammals depicted graphically on the computer monitor. The student must decide whether to flee, fight, or feed. A wrong decision can result in the student's fish being gobbled up by a strong predator. This active participation in a food chain is an excellent way to learn trophic level concepts.

Some simulation programs help teach marine science concepts. *TAG* by M.E.C.C. allows students to capture, tag, release and recapture fish, then use the capture data to calculate population levels.

Pond Ecology by Albion Software allows students to manipulate population levels of inhabitants of a pond community, then observe the effect on other members of the pond ecosystem.

Eutroph by Education Images lets students analyze data from laboratory tests to determine the degree of eutrophication of a still water environment. There are other simulation programs that let students talk to dolphins, track whale migrations and pilot a mini-submarine to explore the ocean bottom.

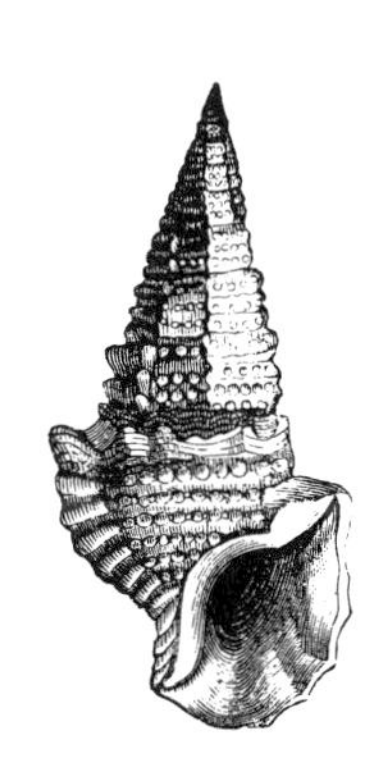

Simulation software allows complex experiments to be conducted quickly and safely right in the classroom. In a one-hour class period a student can electronically run an experiment dozens of times. Much of the number crunching is left to the computer, freeing the science teacher from teaching math and giving students more time to analyze the results and implications of the experiment.

The computer can even generate charts and graphs depicting experimental results. A recent report in *T.H.E. Journal* discussed a meta-analysis of the results of more than 80 computer-related studies. One conclusion of this meta-analysis was that in the teaching of science "science simulations seem to be even more effective than any application in other content areas" (Roblyer 1988).

Microcomputer-based Labs

Microcomputer-based laboratory (MBL) experiments employ specialized devices like thermisters and photo cells to collect data and feed the information directly to the computer.

Science Tool Kit by Brouderbund Software is an inexpensive and easy to use MBL program that can be used in marine studies. The photo cell can monitor fish activity or settling rates of sediments; studies of thermal currents can be enhanced with the thermister. The computer collects the data in an accurate, uninterrupted, unbiased manner, then stores it in memory, displays it graphically, prints it out or allows the student to review the information. Students have more class time to analyze good data rather than collecting questionable data that is tough to analyze.

These remote sensing computer programs also give students a good feel for the new and emerging technologies used by scientists today.

General Application Software

General application software encompasses word processing, data base and spreadsheet programs that are commonly used in a business setting. Teachers familiar with these programs can find applications in the marine science classroom. Spreadsheets can be used to set up "water budgets" to study water conservation. A group of junior high students in Delaware has been conducting weekly physical and chemical tests on several rivers for more than two years, maintaining the information in a computer data base. The computerized record is so easy to manipulate that scientists in the area have sought the students out for consultation on water quality problems. Students can use word processing programs for lab reports or research papers.

There are three possible sources of good software: you can write it, you can buy it or you can locate sources of free, public domain software. Locating and reviewing computer programs to filter out the ones that meet your needs can be a time consuming task. The Computer Education Committee of the Mid-Atlantic Marine Education Association has been compiling information on computer software for several years. If you would like further information on using computers to teach marine science or an annotated bibliography of commercial and public domain marine education software write to: Mid-Atlantic Marine Education Association, Virginia Institute of Marine Science, Advisory Services, Gloucester Point, VA 23062.

References

Barba, B. (1988, November). Computers a la cart. *The Science Teacher*, 44-45.

Kowalski, L. (1984-85, Winter). About the definition of simulations. *Journal of Computers in Mathematics and Science Teaching*, *4*(2) 50-51.

Klemm, B. (1988). Updating the vision for marine education. *Current, the Journal of Marine Education*, *8*(3), 4-9.

McLamb, L.W. & Walton, S. (1985, Fall). Using computers in the marine science classroom. *Journal of Computers in Mathematics and Science Teaching*, *5*(1), 38-43.

McLamb, L.W. & Walton, S. (1987, February-March). Energy relationships in aquatic environments: A computer approach. *Science Activities*, *24*(1), 14-17.

McLamb, L. W. & Walton, S. (1986, February). Getting your feet wet in Iowa. *The Computing Teacher*, *13*(5), 25-26.

Roblyer, M.D. (1988, September). The effectiveness of microcomputers in education: A review of the research from 1980-1987. *T.H.E. Journal*, *86*, 85-87.

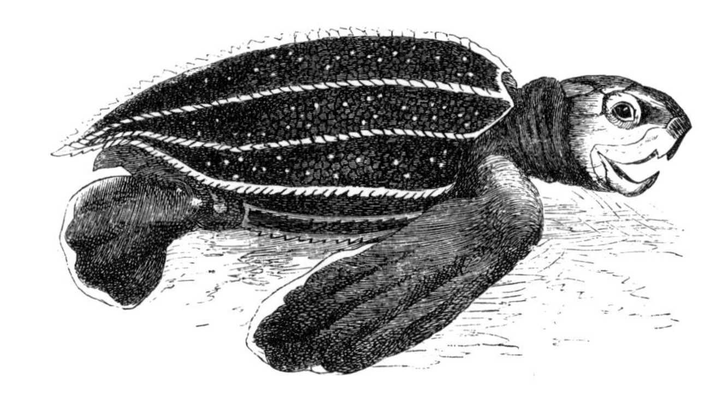

Marine Science Programs Across the Nation

Lundie Spence

Marine science curricula can stretch the walls of your classroom. Imagine the sweet tang of salt air across a marsh, the pulsing roar of wave sets or the sight of thousands of scurrying fiddler crabs. Film and video can bring the sights and sounds, but not the smell; that takes a field trip. Good curriculum materials can guide your lessons to this watery world.

The coastal issues concerning management of erosion, fisheries, water quality, estuaries and wildlife are important to all parts of the United States. The creeks and rivers near you are part of a watershed that leads to the sea. That simple fact ties marine and aquatic programs together and emphasizes the importance of including them in science education.

New resources–lessons, activities, software, films and videos–are being produced in many states. Finding these materials can be a problem, but not an insurmountable one. Excellent materials exist, and this article will help guide you to them.

Marine education curricula have been developed and distributed through three general sources:

1. broad-based curriculum developed for national distribution
2. institutionally developed programs with a long-term commitment for both national and local distribution
3. locally or regionally targeted programs

Broad-based Curricula

The easiest curricula to obtain are those which have a broad-based national distribution strategy or are commercially marketed. Examples are *Outdoor Biological Instructional Strategies (OBIS)* and *Project WILD Aquatic.*

OBIS originated from a 1972 National Science Foundation grant to the Lawrence Hall of Science at the University of California–Berkeley. The present edition of *OBIS* continues the tradition of hands-on activities covering both aquatic and terrestrial concepts. Games, simulations, arts and crafts, and investigations are the basic techniques that lead elementary and middle-grade students to an understanding of biological ideas. The format is well designed for easy application, and the materials are simple.

OBIS modules cover aquatic animal behavior, ponds and lakes, streams and rivers, breakwaters and bays, human impact and the seashore. And if the idea is good, a few slight modifications usually are all that is needed to make the activity perfect for lesson needs. For example, the module "The Food Chain Game," a controlled tag game with popcorn for food and sandwich-bag stomachs, is introduced with terrestrial examples–grasshoppers, frogs and hawks. Why not use shrimp, flounder and sharks instead? *OBIS* activities can easily be modified for students above and below middle grades.

Project WILD (1985) is an interdisciplinary, supplementary environmental and conservation education program emphasizing wildlife. The goal is to assist learners of any age develop the awareness, knowledge, skills and commitment to make informed decisions, behave responsibly and take constructive action. *Project WILD* is usually managed through state agencies of natural resources. Distribution is through a designated in-service setting with trained facilitators.

More than 40 states are involved with *Project WILD* materials. The publications include a secondary and elementary version. The *Aquatic Section* (1987) has a rich selection of marine and freshwater activities. Two favorites are "Turtle Hurdles," which concerns the plight of the threatened loggerhead sea turtle, and "Marsh Munchers," a tag game based on recognition by animal behaviors. *Project WILD* materials are designed to be incorporated into science curricula. All can be related to contemporary issues of environmental concern, which can lead to class discussions in any grade.

Long-term Institutional Developed Curricula

The second source of curricula materials are the well established, institutionally developed curricula. The *Hawaii Marine Science Studies* from the University of Hawaii, and *For Sea,* developed by the Marine Science Center in Poulsbo, Washington, are excellent examples. Both programs are presented to teachers through in-service settings.

HMSS is available in hardcover textbook editions with student workbooks. Two books, *The Living Ocean: Biological Science and Technology of Marine Science* and *The Fluid Earth: Physical Science and Technology of the Marine Environment,* have excellent laboratory exercises fully integrated with content using the inquiry method. *HMSS* is sequentially organized for a one-year course. It has been used successfully on the East and West coasts and inland.

For Sea, a National Dissemination Network curriculum, has curriculum materials for grades 1-2, 3-4, 5-6, 7-8 and 9-12. The units, "Marine Biology and Oceanography–Grades 7 and 8," received recognition in 1983 from NSTA and in 1986 from the

Search for Excellence in Science Education. These materials are teacher resources for modular insertion. Teachers are using *For Sea* materials in 19 states, including Kentucky, Idaho, Kansas and Nevada.

Regionally Targeted, but Useful Nationally

The third general source includes locally or regionally developed materials that often provide the flavor of the area as well as different teaching styles. These vary from county projects like *Project CAPE* in North Carolina to 4-H marine awareness modules to conservation publications. In this group of materials, some are almost ephemeral–one or two printings–while others have maintained stocks for many years.

A new resource is the National Estuarine Research Reserve System, located on all coasts and the Great Lakes. System staff are beginning to design education materials. For example, the North Carolina Division of Coastal Management has produced a curriculum, *Project Estuary* (Jones 1989); a video; posters and a guidebook, *Exploring the North Carolina National Estuarine Research Reserve*. Contact the national office in Washington, DC, for more information on other states.

Living in Water, from the National Aquarium in Baltimore, uses 36 hands-on activities and experiments to teach physical and biological characteristics of marine and aquatic habitats. The curriculum was funded by the National Science Foundation for grades 4 to 6 and received an award from the American Association of Zoological Parks and Aquariums. It has been adopted by Baltimore, Maryland's city school system for all sixth grade students. Teachers from 45 states have requested these materials. The activities include an investigation of water chemistry using chemical testing kits. There are also experiments with dissolved oxygen comparing salt and fresh water, warm and cold water. These are particularly interesting since they relate well to pollution problems with algae blooms and organic dumping into rivers and estuaries. *Living in Water* was developed by teachers under the supervision of the education staff at the National Aquarium.

Sea Grant programs from 30 states have initiated and maintained top quality marine education publications and inservice programs. You can request materials from all coastal states, including Alaska, Hawaii, the Pacific Coast, the Gulf of Mexico, the Atlantic Coast and Puerto Rico. The Great Lakes are represented by New York, Ohio, Michigan, Minnesota, Illinois-Indiana and Wisconsin. Most Sea Grant programs produce monthly newsletters that can help you keep abreast of research, and marine issues. For example, the University of North Carolina Sea Grant's *Coastwatch* has produced issues on the greenhouse effect and red tide, a toxic algae bloom.

Marine Education: A Bibliography of Educational Materials Available from the Nation's Sea Grant College Programs (1988) provides the best source of Sea Grant materials. It is a smorgasbord for all grade levels and all disciplines.

The University of Delaware Sea Grant Program has slide shows for loan or purchase on marine careers, Blue crabs and on walking Delaware beaches. The University of Georgia Extension's new 28-minute video "The Coast of Georgia, Land, Sea and Marshes" can almost bring the taste of salt air into your classroom. In addition, the Georgia Marine Extension Service Library has collected and compiled most of the marine education curriculum available throughout the country. To access this depository, contact Jay Calkins, P.O. Box 13687, Savannah, GA 31416; 312/356-2496.

Wet and Wild, from the University of Southern California, is a Spanish bilingual, multidisciplinary curriculum for kindergarten through sixth grade that reflects some of the issues of the urban coast.

Is Our Food Future in the Sea?, a unit from the Northern New England Marine Education Project, focuses on northern aspects of sea farming such animals as Blue mussels, lobsters and oysters.

Man and the Gulf of Mexico, a four-volume set developed by Mississippi-Alabama Sea Grant, is on the state textbook list in Mississippi and is used extensively in Alabama and Louisiana.

S.E.A. Lab: Science Experiments and Activities for High School Physics, Chemistry and Biology (1990) is a project cosponsored by the University of North Carolina Math and Science Education Center, UNC-Chapel Hill Marine Science Curriculum and UNC Sea Grant College Program. It complements the existing primary (*Coastal Capers*) and middle grade (*Marine Education Materials*) publications from that state.

Alaska Sea Week resources include a fine overview of whales and other mammals in *Mammals and Marine Issues*. The new oil spill curriculum is particularly pertinent.

Ohio has a whole series of publications–*Oceanic Education Activities for Great Lakes Schools (OEAGLS)*–that can guide teachers into biological and environmental lake issues.

A huge mosaic of marine education materials awaits your request. As an inland school teacher, you can flavor your presentations with examples from all our coasts, choose a pertinent activity to illustrate a concept or adopt one curriculum for a year. Issues, research and curricula concerning the marine environment are interdisciplinary. You can integrate chemistry and physics into biology or integrate biology to social issues.

The resources are available; the choice is yours.

References

Aquatic—Project WILD. (1987). Western Regional Environmental Council, P.O. 10860, Boulder, CO 80302-8060.

The coast of Georgia, land, sea and marshes (color video). (1988). University of Georgia Extension, Skiddaway Institute of Oceanography, Box 13687, Savannah, GA 31416.

Coastwatch. UNC Sea Grant, Box 8605, NCSU, Raleigh, NC 27695-8605.

Exploring North Carolina national estuarine reserve. (1988). North Carolina Division of Coastal Management, Box 27687, Raleigh, NC 27611-7687.

For sea. (1986). Marine Science Center, 17771 Fjord Drive, NE, Poulsbo, WA 98370.

Butzow, J., et al. (1984). *Is our food future in the sea?* University of Maine, Sea Grant Advisory Program, 30 Coburn Hall, Orono, ME 04469.

Klemm, B. (Ed.). (1990). *Hawaii Marine Science Studies: The living ocean: Biological science andtechnology of marine science*. HMSS Project, Curriculum Research and Development Group, University of Hawaii, 1776 University Ave., Honolulu, HI 96822.

(1989).*The fluid earth: Physical science and technology of the marine environment*. HMSS Project, Curriculum Research and Development Group, University of Hawaii, 1776 University Ave., Honolulu, HI 96822.

Living in water. (1987). National Aquarium in Baltimore, Education Department, Pier 3, 501 E. Pratt St., Baltimore, MD 21202.

Man and the Gulf of Mexico educational series: Marine and estuarine ecology; marine habitats; diversity of marine plants; marine animals. (1982). University Press of Mississippi, 3825 Ridgewood Road, Jackson, MS 39211.

Marine careers; The blue crab: Beachwalk. (1988). Slide/tape programs. University of Delaware, Sea Grant Advisory Service, 700 Pilottown Road, Lewes, DE 19958.

Marine education: A bibliography of educational materials available from the nation's sea grant college programs [TAMU-SG-401(r)]. (1988). Marine Information Service, Sea Grant College Program, Texas A&M University, College Station, TX 77843-4115.

National Estuarine Research Reserve. Marine and Estuarine Management Division, Office of Ocean and Coastal Resource Management, NOS/NOAA, 1825 Connecticut Avenue, NW, Washington, DC 20236.

Outdoor biological instructional strategies (OBIS). (Request information). Delta Education, Box M, Nashua, NH 03061-6012.

Project Cape resources. (1981-82). Dare County Board of Education, P.O. Box 640, Manteo, NC 27954.

Project WILD (elementary and secondary editions). (1983). Western Regional Environmental Council, P.O. 10860, Boulder, CO 80302-8060.

S.E.A. lab. (1990). 250 pages of high school science activities. UNC Sea Grant, Box 8605, NCSU, Raleigh, NC 27695.

Wet and wild. (1983). [Set of 6 volumes or Unit 1: The physical ocean; Unit 2: Ocean management; Unit 3: Research; Unit 4: Biological ocean; Unit 5: Economic sea; Unit 6: Marine ecology.] USC Sea Grant, Publication Office, University Park, Los Angeles, CA 90089-1231.

Raising Sea-Consciousness in a Landlocked Library

Nancy S. Cowal

Many of the resources needed for marine studies projects are right down the hall, in the library media center. If you think of your school as an ecosystem, you will find the librarian/media specialist to be among the main producers in this web of learning. Not only can this person provide background information and materials to use with your students, she or he can be your team teacher. If you have an idea for a unit, you also have a person to help you follow through.

You can extend the learning environment beyond the classroom or lab by explaining your curriculum needs to the librarian/media specialist. With content and process goals in mind, together you can plan and design instruction that will help your students gain a better understanding of the global importance of our oceans. The librarian/media specialist can let you know what media is already available in your school and can help you obtain other needed information. Use marine studies as a jumping off point and, with the librarian/media specialist's assistance, help students learn how to use research tools, access online information, create computer databases, analyze and apply information and produce their own materials.

Here are a few suggested joint activities:

1. Set up a saltwater aquarium in the library media center where everyone can see it. Or, set up two and pollute one.
2. Make use of an exhibit area. Show marine-related science fair projects. Set up some hands-on activities using any classroom projects (the line may be drawn at fish dissections). Some suggestions: microscopes to view grains of sand from various parts of the world or an oil spill simulation in pans of water.
3. Share and analyze comparative data with a coastal school, either by mail or modem. Design a computer database to collect bird migration or weather information.
4. Exchange natural materials with a coastal school. One class in Cape Hatteras, North Carolina, sent off seashells to a mountain school and received rocks and arrowheads in return.
5. Design and produce books or videotapes to raise the sea-

Flow of Energy Between Classroom & Library Media Center

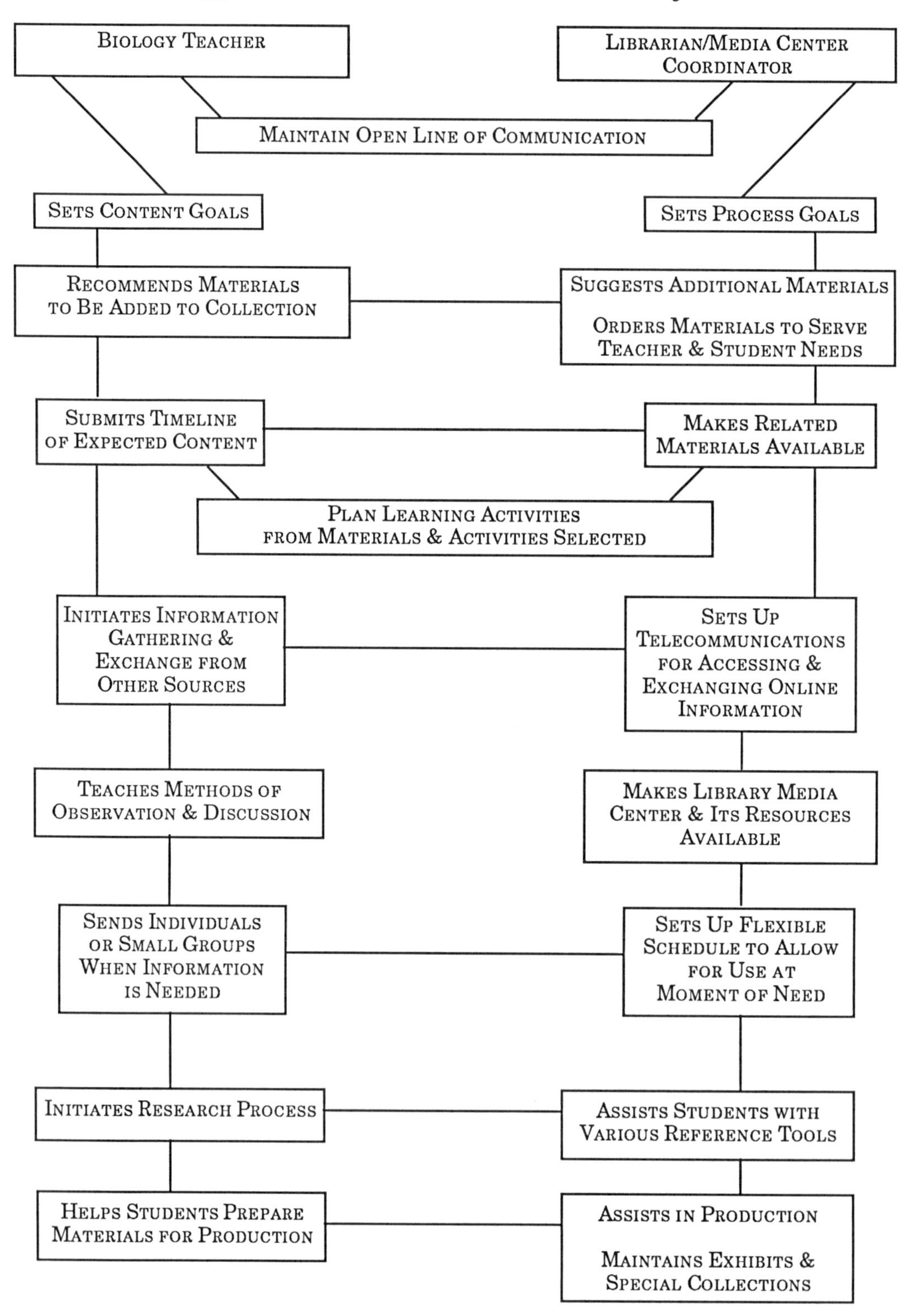

consciousness in your high school or in nearby middle and elementary schools. Let students have a taste of publishing and/or television production.

6. Aim students toward marine science projects for the science fair and enter the best of them in the National Marine Educators' Association World of Water competition. The library media center can be the center for your students' review of literature, experimental data management, graphic representation of results and exhibit design.
7. Assign a good old-fashioned research paper. You'll find there's a new twist to obtaining information. The librarian/media specialist can not only assist your students in using standard print reference materials, but can introduce them to computer-driven sources that can provide hundreds of passages by referring to the keyword "Ocean" from a CD-ROM. Students also can learn to search for a periodical article on marine pollution from an online information service or can "talk" about whale populations with a environmental education professor via an electronic network.

In teaching, the biology teacher and the librarian/media specialist take similar paths: a hands-on, process-oriented approach with an emphasis on information retrieval skills, how the natural world relates to the human cultural world and an understanding of global issues. Go to the library media center and make the connection.

Now for the resources. Listed below are media I have found useful for information, illustration and identification.

Books

Identification guides/natural history

These identification guides and natural history descriptions range from worldwide to specific to the North Carolina coast.

Amos, W.H. (1980). *Wildlife of the islands.* New York: Harry N. Abrams, Inc. Publishers.

Attenborough, D. (1979). *Life on earth: A natural history.* Boston: Little, Brown and Co.

Boschung, H.T., Jr., Williams, J.D., Gotshall, D.W., Caldwell, D.K. & Caldwell, M.C. (1983). *The Audubon Society field guide to North American fishes, whales, and dolphins.* New York: Alfred A. Knopf.

Carr, A. (1986). *A natural history of turtles: So excellent a fishe.* Austin,TX: University of Texas Press.

Coulombe, D.A. (1984). *The seaside naturalist: A guide to nature study at the seashore.* Englewood Cliffs, NJ: Prentice Hall.

Gosner, K.L. (1978). *A field guide to the Atlantic seashore.* Boston: Houghton Mifflin Co.

Harrison, P. (1985). *Seabirds: An identification guide.* Boston: Houghton Mifflin Co.

Hayman, P., Marchant, J. & Prater, T. (1986). *Shorebirds: An identification guide to the waders of the world.* Boston: Houghton

Mifflin Co.
Hoyt, E. (1984). *The whale watcher's handbook.* New York: Doubleday & Co.
Kaplan, E.H. (1982). *A field guide to coral reefs of the Caribbean and Florida.* Boston: Houghton Mifflin Co.
Lippson, A.J. & Lippson, R.L. (1984). *Life in the Chesapeake Bay.* Baltimore: The Johns Hopkins University Press.
McClane, A.J. (1974). *McClane's new standard fishing encyclopedia.* New York: Holt, Rinehart and Winston.
Meinkoth, N.A. (1981). *The Audubon Society field guide to North American seashore creatures.* New York: Alfred A. Knopf.
Minasian, S.M., Balcomb, K.C., III & Foster, L. (1984). *The world's whales.* Washington, DC: Smithsonian Books.
Morris, P.A. (1975). *A field guide to shells of the Atlantic and Gulf Coasts and the West Indies.* Boston: Houghton Mifflin Co.
Randal, J.E. (1968). *Caribbean reed fishes.* Neptune City, NJ: T.F.H. Publications.
Rehder, H.A. (1981). *The Audubon Society field guide to North American seashells.* New York: Alfred A. Knopf.
Silberhorn, G.M. (1976). *Tidal wetland plants of Virginia.* Gloucester Point, VA: Virginia Institute of Marine Science.
Spitsbergen, J.M. (1980). *Seacoast life: An ecological guide to natural seashore communities in North Carolina.* Chapel Hill, NC: The University of North Carolina Press.
Walker, T.J. (1979). *Whale primer: With special attention to the California gray whale.* San Diego: Cabrillo Historical Association.
Walls, J.G. (1975). *Fishes of the northern Gulf of Mexico.* Neptune City, NJ: T.F.H. Publications, Inc. Ltd.

Marine topics

These books help teach marine topics and/or promote understanding of marine issues.

Bascom, W. (1980). *Waves and beaches: The dynamics of the ocean surface.* New York: Doubleday & Co.
Bates, M. (1960). *The forest and the sea: A look at the economy of nature and the ecology of man.* New York: Random House.
Bright, M. (1988). *The dying sea.* New York: Gloucester Press.
Carson, R. (1979). *The edge of the sea.* New York: Houghton Mifflin Co.
Carson, R. (1961). *The sea around us.* New York: Oxford University Press.
Cousteau, J.-Y. (1963). *The living sea.* New York: Harper & Row, Publishers.
Cousteau, J.-Y. (1985). *The ocean world.* New York: Harry N. Abrams, Inc., Publishers.
DeBlieu, J. (1987). *Hatteras journal.* Golden, CO: Fulcrum.
Halle, L.J. (1973). *The sea and the ice.* Boston: Houghton Mifflin Co.

Kraus, E.J.W. (1988). *A guide to ocean dune plants common to North Carolina*. Chapel Hill, NC: University of North Carolina Press.
Lambert, D. & McConnell, A. (1985). *Seas & oceans*. New York: Facts on File.
Kaufman, W. & Pilkey, O.H., Jr. (1983). *The beaches are moving: The drowning of America's shoreline*. Durham, NC: Duke University Press.
Schoenbaum, T.J. (1982). *Islands, capes and sounds: The North Carolina coast*. Winston-Salem, NC: John F. Blair, Publisher.
Simon, A. (1984). *Neptune's revenge: The ocean of tomorrow*. Danbury, CT: Watts, Franklin.
Steinbeck, J. (1986). *The log from the Sea of Cortez*. New York: Penguin Books.
Sumich, J.L. (1980). *An introduction to the biology of marine life*. Dubuque, IA: Wm. C. Brown Company Publishers.
Teal, J. & Teal, M. (1974). *Life and death of the salt marsh*. New York: Ballantine Books.
Trefil, J. (1985). *A scientist at the seashore*. New York: Charles Scribner's Sons.
Zann, L.P. (1980). *Living together in the sea*. Neptune City, NJ: T.F.H. Publications, Inc., Ltd.

Fiction

Do you assign science-related fiction reading? Try some of these titles. Some are the old standbys and others are newer, young adult novels.

Adkins, J. (1983). *A storm without rain: A novel in time*. Boston: Little, Brown and Co.
Cooper, S. (1984). *Seaward*. New York: Atheneum.
Dana, R.H. Jr. (1981). *Two years before the mast*. Cutchogue, NY: Buccaneer Books.
De Hartog, J. (1984). *Star of peace: A novel of the sea*. New York: Harper and Row, Publishers.
L'Engle, M. (1986). *Arm of the starfish*. New York: Dell Publishing Co.
L'Engle, M. (1980). *A ring of endless light*. New York: Dell Publishing Co.
Hemingway, E. (1952, 1980). *The old man and the sea*. New York: Macmillan Publishing Co.
Henry, M. (1947, 1975). *Misty of Chincoteague*. New York: Rand McNally & Co.
Kipling, R. *Captains courageous*. Mattituck, NY: Amereon Ltd.
London, J. (1903, 1931). *The sea wolf*. New York: Bantam Books.
Melville, H. (1981). *Moby Dick* or *The whale*. Berkeley, CA: University of California Press.
Nordhoff, C. & Hall, J.N. (1932, 1960). *Mutiny on the Bounty*. Boston: Little, Brown and Co.
O'Dell, S. (1960). *Island of the blue dolphins*. New York: Dell Publishing Co.

Paterson, K. (1980). *Jacob have I loved.* New York: Avon Books.
Pohl, F. & Williamson, J. (1954). *Undersea city.* New York: Ballantine Books.
Pohl, F. & Williamson, J. (1956). *Undersea fleet.* New York: Ballantine Books.
Pohl, F. & Williamson, J. (1958). *Undersea quest.* New York: Ballantine Books.
Renault, M. (1962). *The bull from the sea.* New York: Random House.
Shute, N. (1957). *On the beach.* New York: Ballantine Books.
Taylor, T. (1969). *The cay.* New York: Avon Books.
Verne, J. *Twenty thousand leagues under the sea.* New York: Bantam Books.
Voight, C. (1984). *Dicey's song.* New York: Atheneum.

Periodicals

These periodicals deal specifically with marine subjects. All review new books and other media and provide information about trips and institutes.

Current is the publication of the National Marine Educators' Association. There are discussions of marine education issues, teaching ideas and news of the organization.

Oceanus, the International Magazine of Marine Science and Policy, is published four times a year by the Woods Hole Oceanographic Institution. It presents in-depth articles describing current research findings.

Sea Frontiers is published by the International Oceanographic Foundation. This bimonthly periodical deals with ocean conservation and natural history topics.

Skin Diver takes you to exotic dive locations all over the world and has up-to-date information on new dive products and diver training.

Audubon, International Wildlife, National Geographic, National Wildlife, Natural History and *Smithsonian* also print many articles on marine subjects. Don't overlook general news magazines and newspapers that give international coverage. The Alaska oil spill and the stranding of three Gray whales in ice were two events that it seemed as if everyone discussed. You can build and sustain interest in and commitment to marine issues by taking advantage of this media coverage. All the above are available to the library media center through a subscription agent.

Maps

Couper, A. (Ed.). (1983). *The Times atlas of the oceans.* New York: Van Nostrand Reinhold Co.

The National Ocean Service produces nautical charts, geophysical maps and tidal current charts. You can obtain free catalogs from: Distribution Branch, National Ocean Service, 6501 Lafayette Avenue, Riverdale, MD 20737; (301) 436-6990

A special map of the ocean floor is available from: Celestial Arts, P.O. Box 7327, Berkeley, CA 94707. The map, which glows in the dark, shows the earth's terrain under the oceans.

Pamphlets/ Other Print Materials

National Sea Grant Depository
maintains a collection of all the publications of Sea Grant programs across the country, as well as search services from a bibliographic database. All are available for loan. National Sea Grant Depository, Pell Library Building, University of Rhode Island, Narragansett, RI 02882.

Sea Grant Abstracts
offers subscriptions free of charge. Each issue lists new publications and gives addresses of all the Sea Grant offices. The materials listed may be purchased from the individual distributing Sea Grant programs. Send a request on school letterhead to: Sea Grant Abstracts, P. O. Box 125, Woods Hole, MA 02543.

There are also bibliographies and lists of other media included in most marine curriculum packages available through Sea Grant programs.

Audiovisuals & Software

National Geographic Society
sells filmstrip kits and videotapes on marine biology, including "The Living Ocean," "A Tidal Flat and its Ecosystem," "Dive to the Edge of Creation," "Riches from the Sea" and "Plankton." They also have a "Whales" videodisk for Level II players. Ask about 30-day approval or their preview policy: National Geographic Society, Department 90, Educational Services, Washington, DC 20036.

Two school television series–"Community of Living Things" and "What On Earth?"–have segments on oceans.

Public Broadcasting System
"Nova," "Nature," "Life on Earth," "The Living Planet" and "The Ocean Realm" have segments on marine life. Videotapes are available for purchase through various library media vendors. Fair use of copyright law entitles you to tape the broadcasts of these programs to use with your classes within 45 days. The librarian/media specialist might be able to do this for you.

Computerized Marine Network
has an annotated list of marine-related computer software and reviews that can be obtained by contacting the Computerized

Marine Network in care of: Susan Walton, 719 Juniper Drive, Newport News, VA 23601. The Network has public domain software on ocean bottom geography and on saltwater fish.

"Voyage of the Mimi"
is a multimedia package on Humpback whale research. It includes print materials, videotapes and computer software. Training sessions for its use are available for school systems that purchase the package. It is designed for grades 6-8, but sections are appropriate for a high school biology class. There also is a stand-alone book available from Bank Street College Project in Science and Mathematics: *The Voyage of the Mimi: The Book.* (1985). New York: Holt, Rinehart and Winston.

Dialog/Classmate
is a computerized online database of educational articles that can be searched and retrieved by students. Call (800) 3-DIALOG for information.

MicroNet
is an online science bulletin board available to North Carolina schools. The supermicrocomputer at Western Carolina University in Cullowhee makes university and state agency resources available to public school teachers and students. Through this service our students and teachers have obtained research materials, assistance with science projects and answers to science-related questions. We have made contact with many schools and have been able to keep joint databases, exchange greetings and information, and even visit one another. The cost is minimal because of grant funding. Check with universities within your state for similar projects.

The Electronic Encyclopedia
is a CD-ROM disk that can give your students instant information. It is available from: Grolier Electronic Publishing Co., Sherman Turnpike, Danbury, CT 06816.

Resources

The Center for Marine Conservation
is an advocate for ocean protection. You can order print and nonprint materials such as a directory of marine education resources, plastics in the ocean environment, songs of Humpback whales on tape, and sea turtle posters. Center for Marine Conservation, 1725 DeSales Street NW, Suite 500, Washington, DC 20036; (202) 429-5609.

Sand collections
are always interesting. Write for a sand donor list and your first sample: Nancy Cowal, Cape Hatteras School Library, P.O. Box 948, Buxton, NC 27920.

Whale adoptions
are another possibility to consider. For information you and your students should write: Whale Adoption Project, International Wildlife Coalition, 634 North Falmouth Highway, P.O. Box 388, North Falmouth, MA 02556.

The librarian/media specialist can acquaint you with various reference tools. Ask for state and department of public instruction film and video catalogs, listings of government publications, and pamphlet catalogs for media you might want to request for loan or purchase. Look in periodical indexes under "Ocean" or "Marine biology" for articles. Check instructional television programming schedules for science series that touch on marine topics. Peruse the library media center's card catalog for books, computer software and audiovisuals under these subject headings: Oceanography, Marine ecology, Marine biology, Marine plants and Marine animals.

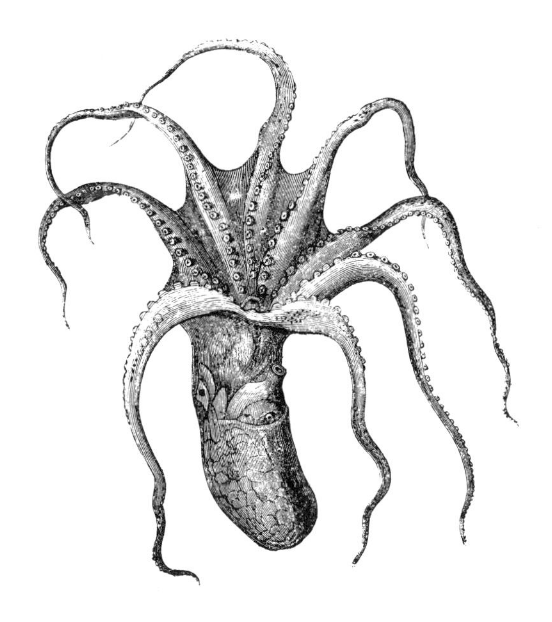

Index